Selected Titles in This Series

Volume

32 **Jean-Charles Grégoire, Gerard J. Holzmann, and Doron A. Peled, Editors**
The SPIN verification system

31 **Neil Immerman and Phokion G. Kolaitis, Editors**
Descriptive complexity and finite models

30 **Sandeep N. Bhatt, Editor**
Parallel Algorithms: Third DIMACS Implementation Challenge

29 **Doron A. Peled, Vaughan R. Pratt, and Gerard J. Holzmann, Editors**
Partial Order Methods in Verification

28 **Larry Finkelstein and William M. Kantor, Editors**
Groups and Computation II

27 **Richard J. Lipton and Eric B. Baum, Editors**
DNA Based Computers

26 **David S. Johnson and Michael A. Trick, Editors**
Cliques, Coloring, and Satisfiability: Second DIMACS Implementation Challenge

25 **Gilbert Baumslag, David Epstein, Robert Gilman, Hamish Short, and Charles Sims, Editors**
Geometric and Computational Perspectives on Infinite Groups

24 **Louis J. Billera, Curtis Greene, Rodica Simion, and Richard P. Stanley, Editors**
Formal Power Series and Algebraic Combinatorics/Séries formelles et combinatoire algébrique, 1994

23 **Panos M. Pardalos, David I. Shalloway, and Guoliang Xue, Editors**
Global Minimization of Nonconvex Energy Functions: Molecular Conformation and Protein Folding

22 **Panos M. Pardalos, Mauricio G. C. Resende, and K. G. Ramakrishnan, Editors**
Parallel Processing of Discrete Optimization Problems

21 **D. Frank Hsu, Arnold L. Rosenberg, and Dominique Sotteau, Editors**
Interconnection Networks and Mapping and Scheduling Parallel Computations

20 **William Cook, László Lovász, and Paul Seymour, Editors**
Combinatorial Optimization

19 **Ingemar J. Cox, Pierre Hansen, and Bela Julesz, Editors**
Partitioning Data Sets

18 **Guy E. Blelloch, K. Mani Chandy, and Suresh Jagannathan, Editors**
Specification of Parallel Algorithms

17 **Eric Sven Ristad, Editor**
Language Computations

16 **Panos M. Pardalos and Henry Wolkowicz, Editors**
Quadratic Assignment and Related Problems

DIMACS

Series in Discrete Mathematics
and Theoretical Computer Science

Volume 32

The SPIN Verification System

The Second Workshop on the
SPIN Verification System
Proceedings of a DIMACS Workshop
August 5, 1996

Jean-Charles Grégoire
Gerard J. Holzmann
Doron A. Peled
Editors

NSF Science and Technology Center
in Discrete Mathematics and Theoretical Computer Science
A consortium of Rutgers University, Princeton University,
AT&T Labs, Bell Labs, and Bellcore

American Mathematical Society

This DIMACS volume contains the proceedings of the second workshop on the SPIN Verification System held at DIMACS on August 5, 1996.

1991 *Mathematics Subject Classification.* Primary 68Q60; Secondary 68Q45, 68Q10, 68M10, 68N99, 68Q68, 68-06.

Library of Congress Cataloging-in-Publication Data

The SPIN Verification System (1996 : New Brunswick, N. J.)

The SPIN Verification System : DIMACS workshop, August 5, 1996 / Jean-Charles Grégoire, Gerard J. Holzmann, Doron A. Peled, editors.

p. cm. — (DIMACS series in discrete mathematics and theoretical computer science, ISSN 1052-1798 ; v. 32)

Workshop held at Rutgers Univ. in New Brunswick, N. J.

Includes bibliographical references.

ISBN 0-8218-0680-7

1. Computer software—Verification—Congresses. 2. SPIN (Computer file)—Congresses. I. Grégoire, Jean-Charles, 1960– . II. Holzmann, Gerard J., 1951– . III. Peled, Doron, 1962– . IV. Title. V. Series.

QA76.76.V47W677 1996

005.2′76—dc21

96-54839

CIP

---

10 9 8 7 6 5 4 3 2 1 02 01 00 99 98 97

CONTENTS

Foreword

The Second SPIN Workshop, held at Rutgers University on August 5, 1996, was part of DIMACS Special Year on Logic and Algorithms. We would like to express our appreciation to Jean-Charles Grégoire, Gerard J. Holzmann, and Doron A. Peled for their efforts to organize and plan the workshop.

The workshop was part of the broader Special Year on Logic and Algorithms program which focused on computer aided verification, finite models, and proof complexity. The special year encouraged collaborations among very different research communities and this volume records one of many workshops in which this was achieved. We extend our thanks to Eric Allender, Robert Kurshan, and Moshe Vardi for their work over many months as special year organizers.

DIMACS gratefully acknowledges the generous support that makes these programs possible. The National Science Foundation, through its Science and Technology Center program, the New Jersey Commission on Science and Technology, DIMACS partners at Rutgers, Princeton, AT&T Labs, Bell Labs, and Bellcore generously supported the special year. Additional funding from Bell Labs allowed increasing the number of scientists who could participate.

Fred S. Roberts
Director

Bernard Chazelle
Co-Director for Princeton

Stephen R. Mahaney
Associate Director

PREFACE

What Is SPIN? SPIN is a general tool for the specification and formal verification of software for distributed systems. It has been used to detect design errors in a wide range of applications, such as abstract distributed algorithms, data communications protocols, operating systems code, and telephone switching code. The verifier can check for basic correctness properties, such as absence of deadlock and race conditions, logical completeness, or unwarranted assumptions about the relative speeds of processes. It can also check for more subtle, system dependent, correctness properties expressed in the syntax of Linear-time Temporal Logic [**14**]. The tool translates LTL formulae automatically into automata representations [**3**], which can be used in an efficient on-the-fly verification procedure.

Some Background. Work on the construction of automated verification systems of this type started about two decades ago. Among the first to built a fully automated tool based on the reachability analysis of finite state models was Jan Hajek at the Technical University in Eindhoven in The Netherlands [**4**]. Between 1976 and 1978 Hajek's system *Approver* successfully uncovered bugs in a series of published designs for communications protocols. The algorithmic techniques on which the *Approver* system was based, were unfortunately never revealed, and therefore the system could only inspire, but not directly influence related efforts by others.

At approximately the same time, Colin West at the IBM research lab in Rüschlikon, Switzerland, worked on the implementation of a tool for Pitro Zafiropulo's duologue matrix analysis technique. This work quickly lead West to develop his own variant of a verification system [**18**]. The most visible result of this work was a first verification, and the uncovering of design flaws, in the X.21 recommendation from the CCITT (now the ITU) with West's perturbation analysis procedure. The X.21 verification is today frequently used as a litmus test for new verification systems.

The work that ultimately lead to the SPIN verification system started at Bell Labs in 1980. The first incarnation of the system, the verifier PAN, started finding bugs in data-switch control protocols in November 1980. Like today's SPIN, PAN was a general on-the-fly verification system, but unlike SPIN it was restricted to the verification of only basic safety properties. Over a period of ten years [**5, 6, 7, 8, 9**], this tool evolved into a powerful verification system with full model checking capabilities. SPIN was first released for general distribution in late 1990 [**10**], and has continued to evolve. Significant improvements in SPIN's model checking capabilities were the introduction of a partial order reduction method in 1994 [**13, 11**] and a built-in translation algorithm [**3**] for converting LTL formulae into the automata recognized by SPIN's verification engine. The code to the SPIN system is available from `http://netlib.bell-labs.com/netlib/spin/whatispin.html`.

Several other tools with a similar long history exist. Descriptions can be found, e.g., in [**1, 15, 16, 12**].

Tool Characteristics. SPIN has several distinguishing features that make it well suited for addressing verification problems in the general area of concurrent software design, and telecommunications systems engineering:

- The specification language for SPIN is a high-level, *asynchronous* and non-deterministic, guarded command language, that is well suited for specifying software process behaviors, instead of a synchronous notation that would be better suited for specifying hardware circuits.
- The logic used in SPIN is based on the linear-time temporal logic LTL, instead of a branching time temporal logic, such as CTL.
- The verification procedure used in SPIN is based on explicit state enumeration, rather than on a symbolic state representation.
- SPIN uses an on-the-fly verification algorithm [**2**], storing as little information in memory as is strictly necessary for completing the verification task, instead of an offline (or two-pass) verification procedure. No transitions (edges) need be stored with this method, and efficient compression and reduction techniques are available to reduce the memory usage to a minimum without incurring undue overhead, or unpredictable performance.
- SPIN uses a general partial order reduction technique [**13, 11**] to exploit regularities that are common in asynchronous interleaving systems, rather than the BDD representations that are common in hardware verification to achieve the same effects.
- SPIN contains a range of simulation options, with either graphical or textual output, that have proven their value in the pre-verification phase of a design. A simple graphical user interface to the system, called XSPIN, enhances the usability of the tool especially to new or occasional users.

The Spin Workshops. This book contains the proceedings of the second workshop on work related to the SPIN verification system. At the workshop, fourteen research papers were presented by researchers from eight different countries.

The keynote presentation was delivered by Moshe Vardi, Noah Harding Professor and chair of Computer Science at Rice University. Prof. Vardi is one of the primary developers of the automata theoretic framework on which SPIN is founded [**17**]. The viewgraphs of the keynote presentation are available as an on-line document at `http://netlib.bell-labs.com/netlib/spin/ws96/vardi.ps`. One other presentation could not be included in these proceedings as a full paper: work reported by Frank Schneider and Jack Callahan from NASA's Software Verification and Validation Facility, on the verification with SPIN of aspects of a distributed system used in one of NASA's upcoming space missions. We have added a paper to this volume that was also distributed as part of the participants proceedings to the SPIN workshop: an outline for an operational semantics definition of SPIN's specification language PROMELA.

Four tool demos were presented at the workshop:

1. The WHEEL environment for SPIN: an extension targeted to the specification and verification of feature interaction problems (by F.J. Lin, Bellcore, USA).
2. A real-time extension of SPIN (by Stavros Tripakis, Verimag, France).
3. A PROMELA to C translator (by Siegfried Loeffler, Hewlett-Packard, England).

4. RSPIN, an extension for the specification and verification of reactive systems (by Frank Olsen, Centre National d'Etudes des Telecommunications, Issy-Les-Moulineaux, France).

The number of places where the SPIN system is installed and used now numbers in the thousands. The program of this year's workshop reflects the nature of the work in formal verification that is triggered or inspired by this system. In four broad categories, there are:

- Theoretical and foundational studies.
- Empirical studies of the relative effectiveness of different types of search and storage algorithms.
- Significant practical applications.
- Extensions and revisions of the basic SPIN code.

Several of the projects in each category were represented at the workshop, but perhaps more encouraging still is that many more of these SPIN projects are in progress at academic and industrial research labs around the world. The goal of the SPIN workshops is to create an opportunity for those who work with this system to meet, share experiences, learn about each others work, and exchange ideas. As witnessed by these proceedings, this year's workshop fully met that goal.

Acknowledgements. The 1996 workshop was sponsored by Bell Labs, and by DIMACS, the National Science Foundation's Science and Technology Center for Discrete Mathematics and Theoretical Computer Science, as part of their Special Year Program on *Logic and Algorithms*. DIMACS provided both logistic support at the workshop location at Rutgers University in New Brunswick, New Jersey, and provided financial support for invited speakers and graduate students. We are especially grateful to Pat Pravato, Sarah Donnelly, Wanglai Li, and Hangbiao Shi for their courteous and efficient help with the preparations, often under time pressure of preparations for other special year events.

Jean-Charles Grégoire, Gerard J. Holzmann, Doron Peled.

References

[1] E. M. Clarke and E. A. Emerson, Characterizing properties of parallel programs as fixpoints, *Proc. 7th Int. Coll. on Automata, Languages and Programming*, LNCS 85, 1981.

[2] C. Courcoubetis, M. Vardi, P. Wolper, M, Yannakakis, Memory-efficient algorithms for the verification of temporal properties, Formal methods in system design 1 (1992) 275–288.

[3] R. Gerth, D. Peled, M. Vardi, P. Wolper, Simple on-the-fly automatic verification of linear temporal logic, *Proc. 15th Int. Conf. on Protocol Specification, Testing, and Verification*, INWG/IFIP, Eds. P. Dembinsky, and M. Sredniawa, Warsaw Poland, 1995.

[4] J. Hajek, Automatically verified data transfer protocols, *Proc. 4th ICCC*, 1978, Kyoto, pp. 749-756.

[5] G. J. Holzmann, *PAN - a Protocol Specification Analyzer*, Bell Laboratories Technical Memorandum, TM81-11271-5, May 1981.

[6] G. J. Holzmann, R. A. Beukers, The PANDORA protocol development system, *Proc. 3rd Int. Conf on Protocol Specification, Testing, and Verification*, INWG/IFIP, Eds. H. Rudin and C. West, pp. 357–369, North Holland Publ. Co., June, 1983.

[7] G. J. Holzmann, Tracing protocols, *AT&T Techn. Journal*, Vol 64, No. 10, Dec. 1985.

[8] G. J. Holzmann, An improved reachability analysis technique, *Software Practice and Experience*, Vol. 18, No. 2, pp. 137-161, Feb. 1988.

[9] G. J. Holzmann, J. Patti, Validating SDL specifications: An Experiment, *Proc. 9th Int. Conf on Protocol Specification, Testing, and Verification*, INWG/IFIP, Ed. C. Vissers and E. Brinksma, Twente, Neth., June, 1989.

[10] G. J. Holzmann, *Design and Validation of Computer Protocols*, Prentice Hall, 1991.

[11] G. J. Holzmann and D. Peled, An improvement in formal verification, *Proc. 7th Int. Conf. on Formal Description Techniques*, Eds. D. Hogrefe and S. Leue, FORTE94, Berne, Switzerland. October 1994.

[12] R. P. Kurshan, Computer Aided Verification of Coordinating Processes, Princeton University Press, 1994.

[13] D. A. Peled, Combining partial order reductions with on-the-fly model checking, *Proc. 6th Int. Conf. on Computer Aided Verification*, CAV94, Stanford, Ca., June 1994.

[14] A. Pnueli, The temporal logic of programs, *Proc. 18th IEEE Symposium on Foundations of Computer Science*, Providence, R.I., pp. 46-57, 1977.

[15] J. P. Queille, *Le systéme César: description, spécification et analyse des applications réparties*, Ph.D. Thesis, June 1982, Computer Science Dept., Univ. Grenoble, France.

[16] K.L. McMillan, *Symbolic model checking: an approach to the state explosion problem*, Kluwer Academic Publ., 1993.

[17] M.Y. Vardi and P. Wolper, An automata-theoretic approach to automatic program verification, *Proc. First Symposium on Logic in Computer Science*, June 1986, Cambridge, pp. 322–331.

[18] C. H. West, General technique for communications protocol validation, *IBM Journal of Research and Development*, Vol 22, No. 4, p. 393, 1978.

DIMACS Series in Discrete Mathematics
and Theoretical Computer Science
Volume **32**, 1997

State Space Compression with Graph Encoded Sets

J.-Ch. Grégoire

ABSTRACT. We explore the application of a state space compression algorithm to SPIN. Compression techniques are notably unpredictable: we tend to use them when they work well for us (e.g. ZIP), but we tend to forget that, according to the theory, they can also result in an expansion (e.g., ZIP twice).

How well a compression algorithm works in a given context is typically discovered experimentally. In this paper, we describe such an experiment, where a compression algorithm based on sharing of state information in a graph structure is used as a state store for SPIN-generated verification programs.

The positive results are encouraging for further investigation of the use of this technique in SPIN-like tools.

1. Introduction

The amount of main storage available on a given computer is the usual bottleneck in the verification of finite state systems. In SPIN [**H91**], this obstacle is alleviated somewhat in two different ways. The first is through the use of "partial order" techniques [**HP94**], which reduce the size of the system to explore while preserving the properties we want to verify. The second consists of changing the search from an exhaustive one to a partial, stochastic one, but without guaranteing a complete exploration [**H91**]. Those two techniques are orthogonal and can be combined.

We explore here the use of a data structure, called Graph Encoded Sets (GE-SETS) [1] to support the computations of SPIN. This data structure acts as a compression mechanism for the state space. By encoding the state representation, it has the potential of reducing the amount of storage required to hold the state space. However, since it acts as a compression mechanism, it also has the potential of actually using *more* storage than would be required without compression. Such behavior is unavoidable with compression, where a reduction can never be guaranteed. However, we know in practice that some do work in most cases as the Lempel-Ziv

Key words and phrases. 68P05 (data structures), 68U30 (Computer Aided Verification).

[1] The original French name of the data structure was Arbres Partagés, or "shared trees", and dubbed in English "sharing trees" by the original authors. We have found the name to be misleading, and prefer the GE-SETS terminology; GE-SETS, however, are indeed Arbres Partagés.

based (e.g. ZIP) tools bear witness. It is therefore critical to study a compression mechanism in its potential context of utilization.

In earlier work [**GGZ95**], we have already studied GE-SETS in the context of a custom reachability analysis tool. Here our focus is on SPIN, and on whether or not this algorithm will behave favorably (i.e. achieve good compression in reasonable time) in the context of SPIN, and with "partial order" reduced explorations. Our results in this experiment have been largely positive, but we should stress they are only the first step towards a GE-SETS-based SPIN.

Next, the data structure is defined, and its properties are illustrated. We then define a few of its operators. We go on to explore how they can be used in SPIN, and show the results we have obtained in a few experiments. Some observations are made and discussed.

2. Definitions

Let us begin with a formal definition of GE-SETS. Without loss of generality, we consider here a single valuation function for all domains.

A GE-SET is a rooted acyclic graph $(N, V, val, succ)$ with a layered structure such that $N = N_0 \cup \ldots \cup N_k, k \geq 0$, is a finite set called the set of nodes, where N_i is the set of nodes of layer i, V is a set of values, $val : N \longrightarrow V \cup \{\top, \bot\}$ is the valuation function, $succ : N \longrightarrow \mathcal{P}(N)$ [2] gives the successors (i.e. sons, on the next layer) of a node; and the following properties hold:

1. $\forall i \mid 0 \leq i < k, \forall n \in N_i, succ(n) \subseteq N_{i+1}$: each node has all its successors on the next layer;
2. $\forall i \mid 0 \leq i \leq k, \forall n_1, n_2 \in N_i, n_1 \neq n_2, val(n_1) = val(n_2) \Rightarrow succ(n_1) \neq succ(n_2)$: two nodes holding equal values in the same layer do not have the same set of sons;
3. $\forall n \in N, \forall s_1, s_2 \in succ(n), s_1 \neq s_2 \Rightarrow val(s_1) \neq val(s_2)$: a node does not have two sons with equal values;
4. $|N_0| = 1$ and $\forall n \in N, val(n) = \top \Leftrightarrow n \in N_0$: the first layer N_0 contains only one element (called the root), the only one with value $\top$;
5. $val(n) = \bot \Leftrightarrow succ(n) = \emptyset$

Let us now consider that each component of a sequence is mapped to a specific layer, according to the implicit sequence order. The second condition guarantees that there will be *suffix merging* of sequences that share equal ends, while the third condition guarantees the *prefix merging* of sequences that share equal beginnings. This condition furthermore imposes the creation of the longest sharable prefixes, or in other words, that *every prefix is unique*.

Any path starting from the root node $r \in N_0$, with value $\top$, and terminated by a node with value $\bot$ represents one sequence of values encoded in the GE-SET.

The following GE-SET (fig. 1) represents the set $\{(a, b, d, e), (a, b, d, f), (a, b), (a, c), (a, c, d, f)\}$. Note that sequences of different lengths can be represented in the same GE-SET. Notice also the repetition of a node with value d in the 3rd layer, which is legitimate since the set of suffixes is different for the two nodes.

A fundamental property of GE-SETS is the *uniqueness* of the coding: there is only one structure (up to isomorphism) satisfying the properties for any set of sequences. This, however, does not imply *minimality*; it may often be possible to find graph codings of sequences which would have a smaller number of nodes and/or

[2] $\mathcal{P}(N)$ is the *power set* of N.

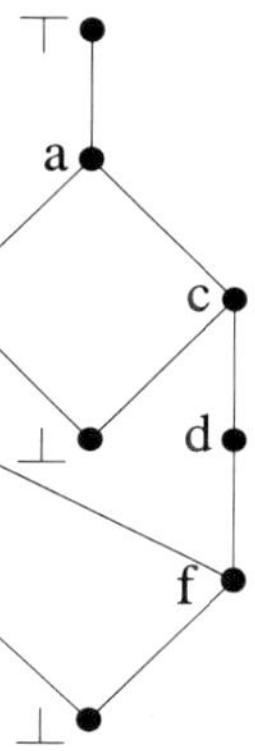

FIGURE 1. Example GETS

vertices. An important corollary of uniqueness is *constant access time*: the time required to check for the presence of any sequence in a GE-SET depends only on the length of the sequences, but not on the number of sequences in the GE-SET. This means that checking for the presence of a sequence in a GE-SETcan be done without backtracking.

The GE-SET uniqueness property derives from the uniqueness of the prefixes (like tries), and the sharing of identical suffixes on the same level. An informal proof follows: imagine [3] two GE-SETS holding the same set of sequences. Let us also imagine that, following a path simultaneously through both GE-SETS, we encounter a node which has a different number of sons in each GE-SET. This would mean that, in one of the two GE-SETS, there would have to be another path leading to (one of) the son(s) missing in the other, since both GE-SETS hold the same information. This however would break the rule of the sharing of the longest prefix. The same reasoning applies to suffixes.

GE-SETS have the following characteristics:

Explicitness:: each contribution to the sequence exists in the graph;

Constant time projection:: the extraction of subsequences on all elements of the set of sequences can be done in constant time;

Heterogeneity:: the domains of the components of the sequences can be different and the sequences can have varying lengths;

Insertion order independence:: since the representation is unique, it does not depend on the order in which the sequences are inserted in the GE-SET;

Isomorphism:: again, because of the uniqueness property, two GE-SETS containing the same set of sequences are isomorphic.

3. Behavior of GE-SETS

The following example will further illustrate the structure of GE-SETS. Consider the following GE-SET holding the set $\{(a, b, b, a), (a, b, b, b), (b, b, b, a), (b, b, b, b)\}$. This structure is represented in figure 2. We add to it the sequence (a, b, b, c). The structure of the graph changes significantly, as it is split to isolate the two longest prefixes (a, b, b) and (b, b, b), to preserve the second property, as shown in figure 3.

[3] A more formal proof is given in [**ZL95**].

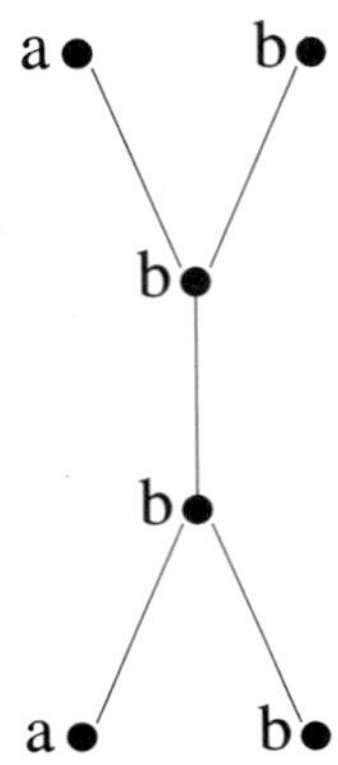

FIGURE 2. Simple GE-SET

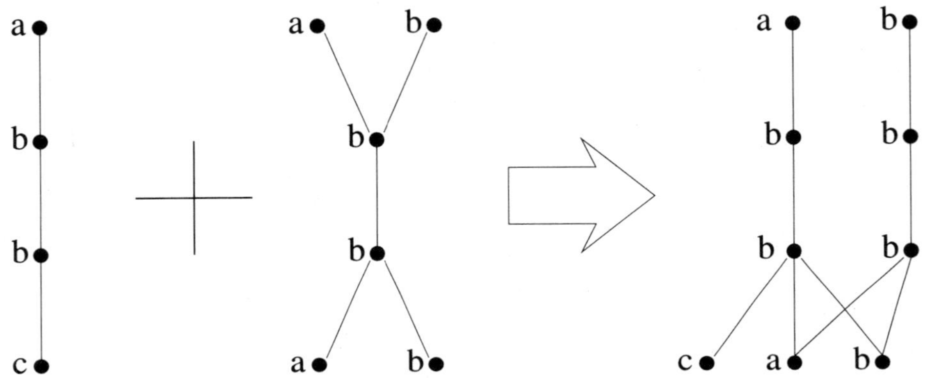

FIGURE 3. First addition

Observe also that this structure is not minimal in the number of node and links. Figure 4 shows an alternate graph holding the same information, with the same number of nodes, but less links. It is however not a GE-SET since the prefix (a, b, b) is duplicated.

Let us now add the sequence (b, b, b, c) to the GE-SET in figure 3. Observe on figure 5 that the result is more compact. The addition of the sequence has allowed *suffix merging*: the creation of a common sharable suffix, its merging and, in this case, the recursive merging of the paths leading to it into a common one.

We shall now introduce a set of operators on GE-SETS.

4. Operations on GE-SETS

We present here a few operators to manipulate GE-SETS. We must however first introduce some terminology, which is derived from graph and language theory. A GE-SET is an undirected graph structure organized along layers, where vertices connect nodes on adjacent layers only. We have already indicated that any path between the top layer of a GE-SET and terminated by the special $\perp$ value corresponds to a sequence of information stored in it. Any path is composed of a *prefix,* a *body* and a *suffix,* any of which might be empty. When we talk about the insertion

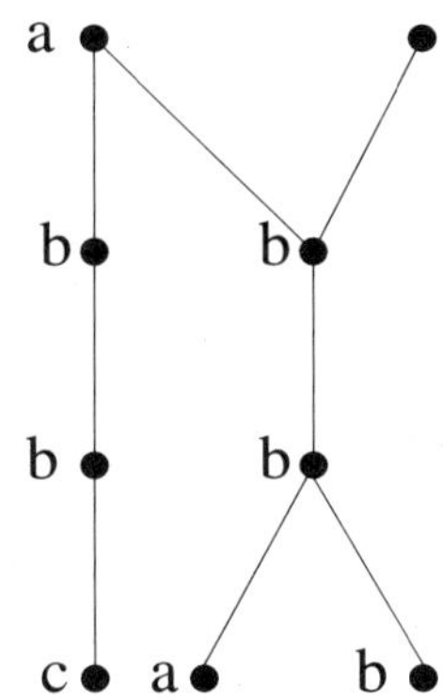

FIGURE 4. Smaller graph, but not a GE-SET

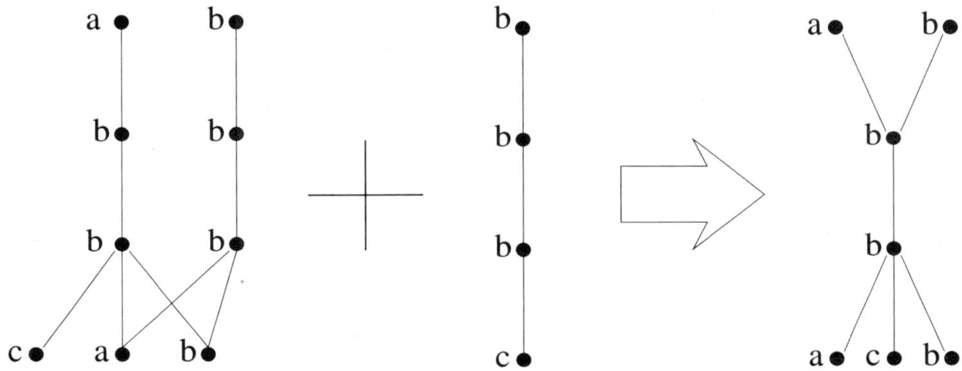

FIGURE 5. Second addition

of a sequence of values into a GE-SET, we will have to take into account existing prefixes and suffixes which can be shared to some extent. A prefix here is the part of a path starting at $\top$ leading to a node with more than one parent. A suffix is the part of the path ending at $\perp$ which has nodes with single sons only. The body is simply the part of the sequence between the prefix and the suffix. We call a node with more than one children a split node, and a node with more than parent a merge node. A node can obviously be both a split node and a merge node.

4.1. Operators. GE-SETS can support all the operators of set manipulation [**ZL95**]. For the purpose of this presentation, however, we present only three basic operators, namely *membership, insertion* and *deletion*. An important issue in the implementation of these operators is whether an insertion or removal operation should preserve the GE-SET, or whether some of the constraints could be relaxed for a while, until some *normalization* operation is performed. In fact, one original implementation of GE-SETS was designed exactly that way. However, our analysis of several tests cases have convinced us that, on any number of repetitions, using an operation preserving the GE-SET resulted in general better performance than a non-GE-SET preserving operation, followed by a normalization [**GGZ95**].

4.2. GE-SET **membership.** GE-SET membership is the simplest operator, and proceeds simply by checking if there is a $\perp$-terminated path in the graph matching a sequence given as parameter. At each layer, we check for the existence of a successor with the appropriate value, and repeat the process from the successor as long as there is a match, until the successor is $\perp$.

4.3. GE-SET **insertion.** GE-SET insertion proceeds as follows.

1. Identify in the GE-SET the longest prefix, the body and the longest suffix matching the sequence of values to insert; the prefix and the body are contiguous, but the suffix may be detached if either the prefix or the body end with $\perp$, then the sequence already exists in the GE-SET and we can terminate, otherwise:
2. If the body isn't empty, then duplicate and split it on its whole length
3. Expand the path from the end of the body (or the prefix if the body is empty, or from $\top$ if both are empty) to the start of the longest suffix or to $\perp$ if the latter is empty, and
4. Merge suffixes upward.

In the previous example (fig. 3), the prefix was (a, b), the body (b, b), and the suffix was empty. The body was duplicated to create a (longest) prefix (a, b, b) in the GE-SET. A new suffix was then added. In the second example (fig. 5), the prefix was (b, b, b), the body was empty, and the suffix was (c). A link was added from the end of the prefix to (c), then the nodes were merged upwards.

Adding a new suffix creates an opportunity for sharing similar suffixes. We must identify, from the node where the path merges with the suffix if another node with the same value and the same set of sons already exists. If it is the case, the nodes can be merged, and we repeat the process upward.

It is quite clear that this algorithm terminates. Once we have completed the upward merge, nothing is left to do since the longest unique prefix constraint was established from the start, and is not invalidated by a merge. We see also that the procedure is linear in the number of layers.

4.4. GE-SET **deletion.** GE-SET deletion operates similarly to GE-SET insertion.

1. Identify a complete (i.e. to $\perp$) path matching the sequence of values we want to remove, terminate if it is not found; otherwise
2. Decompose the path into longest prefix, body and suffix
3. If the body isn't empty, split the path (i.e. expand the prefix) from the beginning of the body to, and including the node where the suffix starts
4. If the suffix is empty, truncate the path from the end up to the first split node, otherwise
5. Truncate the suffix from that node to any merge node and
6. Merge the suffixes upward.

The intuition behind this algorithm is quite simple. We must isolate the tree holding the path to be removed from the effects of sharing, then we remove its distinctive part, preserving the longest prefix distinguishing the sequence to be removed, and normalize the new resulting suffix.

4.5. Other Operators. We have also implemented an *iterator* on GE-SETS. An iterator call successively returns all the sequence instances held in a GE-SET.

Our iterators are robust in the presence of modifications to the GE-SET, but do not guarantee that all the sequences stored will be found in the current sweep in that case.

For our purposes, we do not require operators to merge GE-SETS together or to extract subsets of them. They can however be defined in terms of the operators we have introduced, and could be implemented that way, although it might not be the most efficient way. The same applies to complementation.

5. Application to SPIN and Analysis

To analyze the performance of GE-SETS in a "real" context, we have modified analyzers generated by SPIN (i.e. "pan" programs) so that they would use this data structure rather than the hash table storage. Since this was only a feasibility study, we haven't made any attempt at guaranteing that the substitution was compatible with all the algorithms supported by SPIN in full state search mode. We have focused only on safety properties, with or without partial orders reduction.

We have analysed three examples: two tests from the SPIN distribution, namely *snoopy* and *pftp,* and *gsm,* a development of our group, inspired by one of the GSM protocols.

We first explore the issues we were confronted with, namely *state coding* and *stack presence coding.* We then cover the examples, report the performance measured, make some observations and discuss them.

5.1. State Coding. The most significant difficulty of using GE-SETS is to define an encoding for state information. The most straightforward is, of course, to allocate a layer to every variable in the PROMELA model, as well as the bookkeeping variables of SPIN (e.g. the "process counters" _p).

In our experiments, we have done some clustering of the variables. The simple rule has been to cluster variables in each of the global-queue and process groups to keep the size of the domain of the cluster to less then 2^{16} (a *short word*).

The clusters have been assigned to layers following simple rules: put all the clusters of a group item together and try to put in close proximity the group items interacting together.

Our implementation of GE-SETS uses a static allocation of pointers for each element of the domain of clusters. Having a sparse cluster thus leads to a waste of space. When variables with a sparse domain are combined together in a cluster, the wastefullness is compounded. Independently of the (PROMELA) model itself, transitions identifiers typically have a sparse domain in SPIN. This phenomenon is thus real. However, its importance is relative: the waste in pointer space may or may not not be very significant, depending on the size of the domain, and the degree of sparseness.

How clustering is done is another possible contributor to space or time overhead. The most straightforward form of clustering is to compose the *declared* (i.e. in the model) domains of the variables. PROMELA is however notably deficient in this respect, since declared domains can be much larger than the reachable domains. We have tried instead to use the number of bits required to store the reachable domains. Support for such an idenfication exists in SPIN (`VAR_RANGES`). It could also largely be done by a static analysis of the PROMELA model. In practice, on our examples, this identification has been straightforward [4]. The domains are rounded up to the next power of two and the clusters are assembled through shifting and "or"-ing.

More complex solutions exist which would involve a look-up to compress sparse domains, and clustering by multiplication rather than by shifting, for a most compact clustering. We haven't used these techniques in our experiments.

5.2. Stack Presence Coding. Supporting partial orders has been challenging since the algorithm used in SPIN requires the recognition of whether or not a state already encountered is on the stack. In full state search mode in SPIN, this is done by tagging the state information in the store. The value is the tag is set when the state is met the first time, and reset when it is popped from the stack.

We have found two alternatives to support such a mechanism with GE-SETS. The first, obvious one is to have another GE-SETS to hold a copy of the states currently on the stack. For any state, we would have three GE-SETS operations: addition to the store and to the stack, deletion from the stack. We have however tried another method, borrowing from the idea of tagging the state. We store the state *twice* in the state store, with different tag values appended: one, with a tag value of 0, when the state is first encountered and thus is on the stack, and another, with a tag value of 1, when it is popped from the stack. Except with the REACH compile time option, where the state space is always going to be explored at some specific depth, and we need to memorize the depth information with the state, this is sufficient to quickly decide whether or not a state is on the stack.

The cost of this algorithm is two insertions, rather than one, and a doubling of the amount of states stored in the GE-SETS. However, since we know that both instances of the state will end up being in the store, this doubling practically involves a lot of sharing and doesn't appear to lead to significant space overhead in practice.

5.3. snoopy. We have tested snoopy only in NOREDUCE'd mode. The amount of memory required otherwise is too small to give significant results. The encoding uses 21 layers, the largest domain used, for the `cache` processes is 2560.

	states	memory (Mb)	time (s)
FS, spin	91920	9.934	13.45
FS, GE-SETS	91920	0.55(GE-SETS) + 0.375(spin)	45.52

The memory contribution for GE-SETS is broken down in two parts: GE-SETS *per se* and SPIN which is mostly the memory required for the stack. We have a compression factor specifically for the state space of about 18 in this case.

5.4. pftp. This example being larger, we have been able to test it both in REDUCE'd and in NOREDUCE'd form. We have 19 layers, and the largest domain is 4096. These results are for PFTP with loss, but without duplication.

[4] Encoding and implementation were done in the course of a couple of hours in all cases.

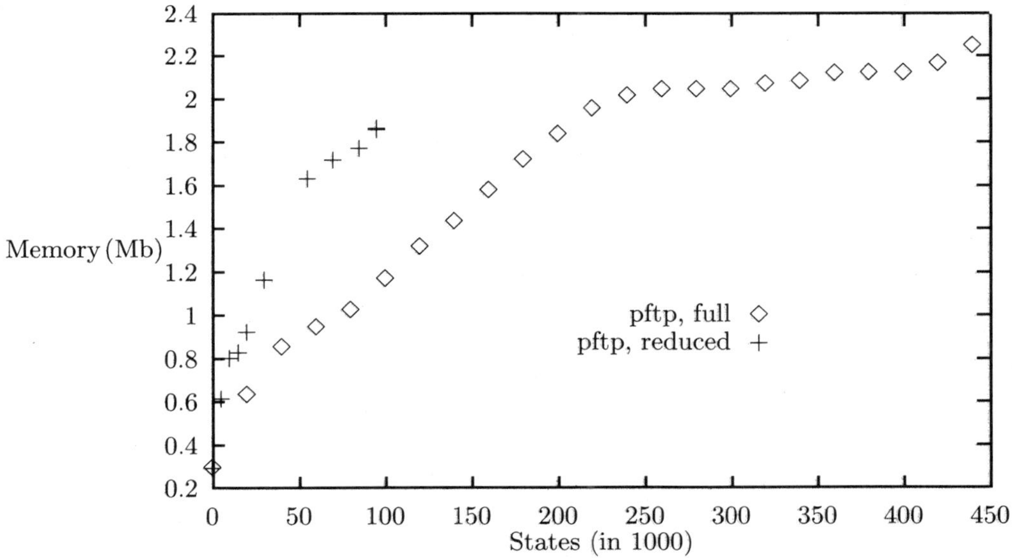

FIGURE 6. pftp, spin & GE-SET, memory behaviour

	states	memory (Mb)	time (s)
FS, spin	439895	54.49	86.71
FS, GE-SETS	439895	2.25(GE-SETS) + 0.334(SPIN)	179.59
RS, spin	95241	13.33	10.41
RS, GE-SETS	95241	1.864(GE-SETS) + 0.334(SPIN)	42

We achieve here a compression rate of 24 for the NOREDUCE'd mode, and about 7 for the REDUCE mode. Note also that, in REDUCE mode, the difference in time performance is a factor of about 4, but we store twice at many states in the store.

The static part of the GE-SETS is only about 135k.

The following figures represent the evolution in the memory and time requirements of the computation as a function of the number of states for the GE-SETS-based examples. The time results are also represented for SPIN: in this case, the use of memory is always linear as a function of the states [5].

The memory consumption behavior represented in figure 6 is quite interesting. The use of memory isn't linear, but presents a bit of a quirk. This means that the use of memory improves at higher capacities in this case. Both the REDUCE and NOREDUCE explorations present the same phenomenon, although it is more distinct in the NOREDUCE case. Predictably, reducing the number of states has removed some opportunities for sharing.

The time behaviors are interestingly enough almost linear in both cases (fig. 7) and also for the SPIN versions (fig. 7). The linearity is somewhat surprising, since we could have expected the quirk to show up in the time measurements. SPIN 's behavior is also linear in terms of the number of states, which means that, in

[5]Variations can be due to the memory used by the stack, but this one tends to be allocated early in the execution.

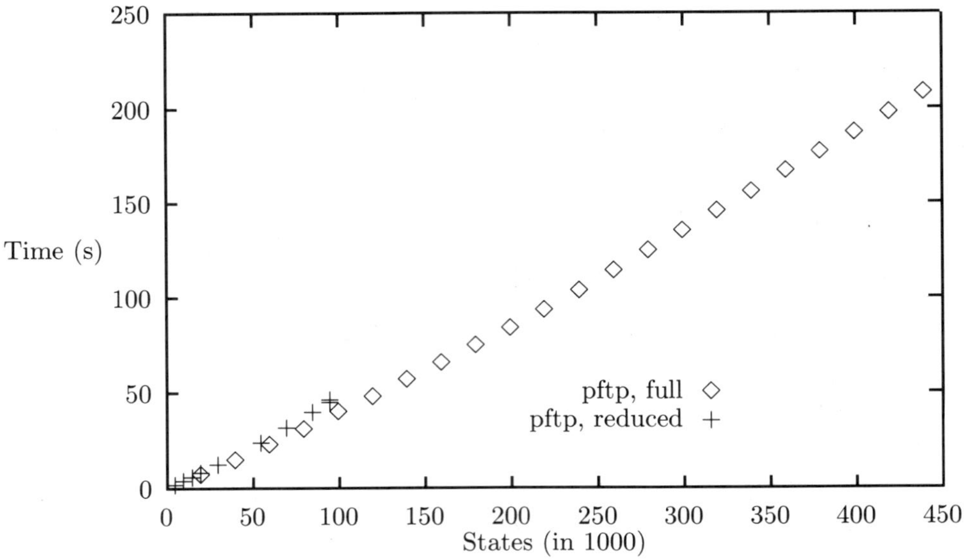

FIGURE 7. pftp, spin & GE-SET, time behaviour

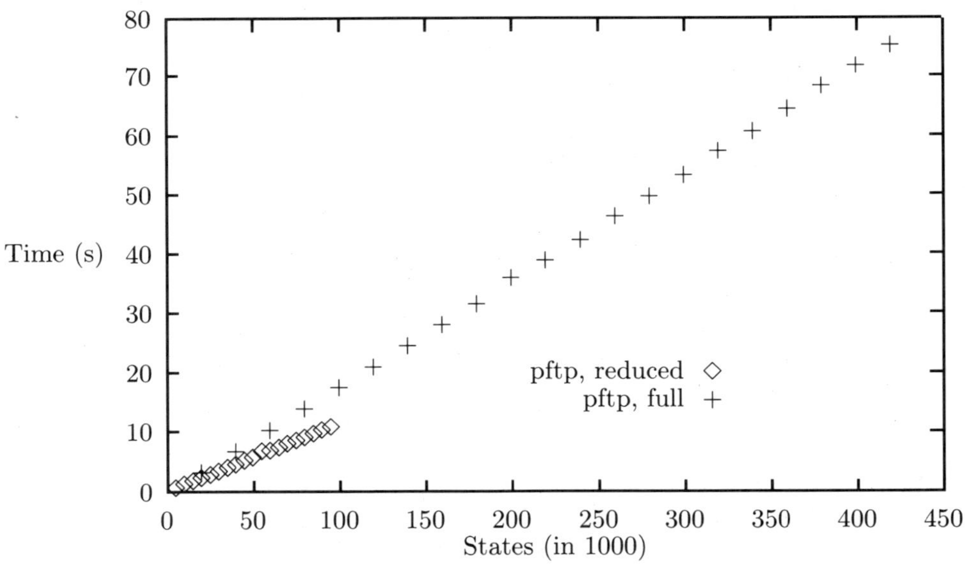

FIGURE 8. pftp, pure spin, time behaviour

general, the rate of discovery of new states tends to remain constant. This rule also applies to the GE-SET version.

5.5. gsm. This last example is not a "classic" of the SPIN distribution. It is a small, simplified model of a call set-up for a mobile in a GSM system. The

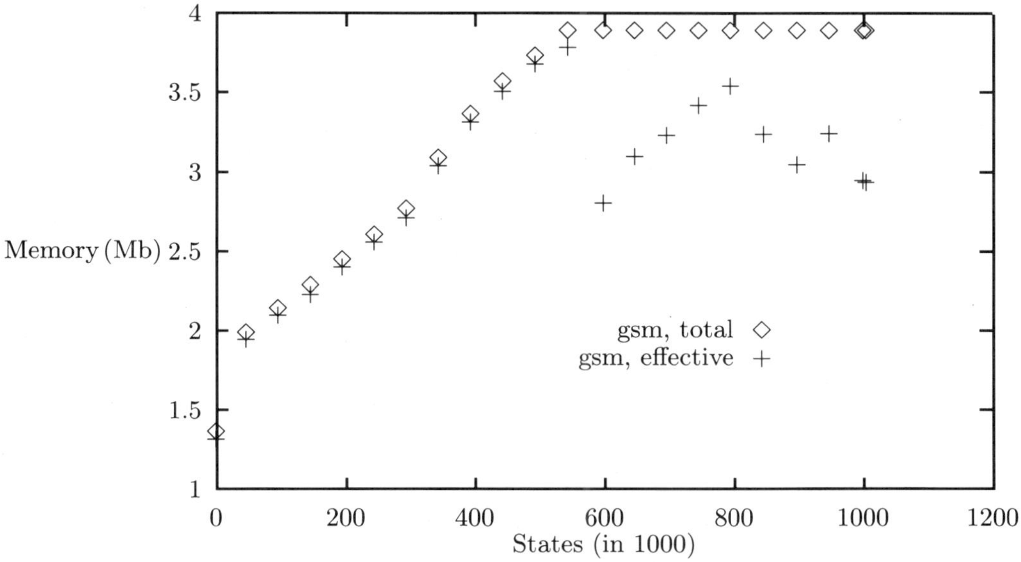

FIGURE 9. gsm, GE-SET, memory behavior

state vector is 144 bytes wide, the maximum depth reached in a REDUCE'd search is 196999. The GE-SETS has 20 layers, and the largest domain size is 16384.

	states	memory (Mb)	time (s)
RS, SPIN	1004630	140.267	251.780
RS, GE-SETS	1004630	4.022(GE-SETS) + 6.4(SPIN)	634.92

We see that the compression factor achieved here was over 33, with an execution time less than the double, and twice the states, since we are in REDUCE mode. The runtime behavior is even more interesting. The memory vs. states graph (fig. 9) shows that halfway through the exploration, the program had allocated all the storage it required, and the rest of the time was spent recycling storage. The wide variations in the use of memory do not however show in the time vs. states graph (fig. 10), which is, once again mostly linear.

For a comparison, a comparable bitstate search, in terms of the amount of storage used, gave the following results [6], after 192s:

```
./pan -m200000 -w25
(Spin Version 2.8.6 -- 17 May 1996)
        + Partial Order Reduction

Bit statespace search for:
        never-claim             - (none specified)
        assertion violations    +
        cycle checks            - (disabled by -DSAFETY)
        invalid endstates       +
```

[6]The depth reached in this case is indeed greater than a full search's; this is not unreasonable, since the greatest depth may be reached through a different path.

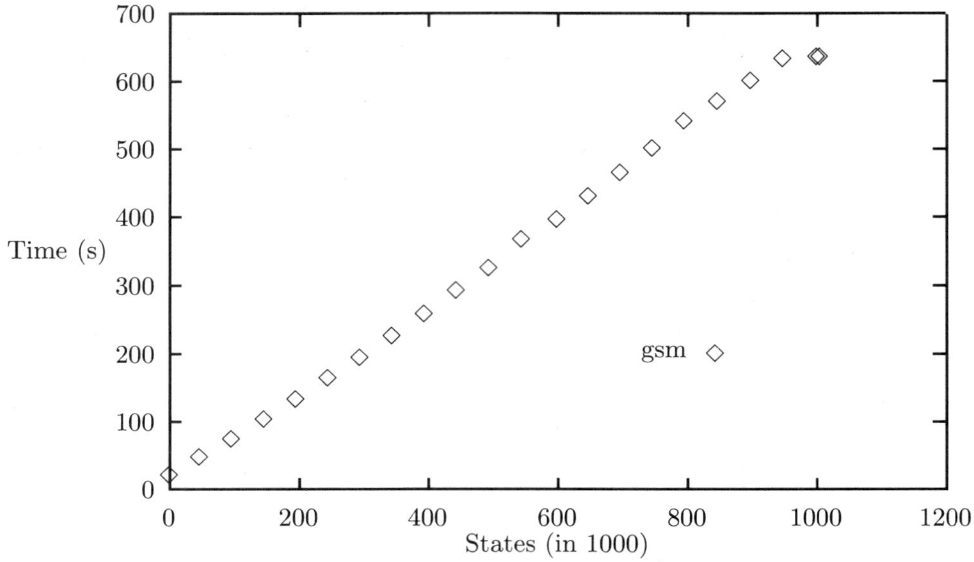

FIGURE 10. gsm, GE-SET, time behaviour

```
State-vector 144 byte, depth reached 197683, errors: 0
1.00088e+06 states, stored
6.15347e+06 states, matched
7.15435e+06 transitions (= stored+matched)
5.56773e+07 atomic steps
hash factor: 33.5248 (expected coverage: >= 98% on avg.)
(max size 2^25 states)

1.49727e+08   equivalent memory usage in bytes (stored*vector + stack)
1.55943e+07   actual memory usage

...
```

We see that the bitstate mode search does indeed give us a fairly good approximation in a good time. It is actually significantly better than the 98% expected coverage (20000 vs. 4000 states) reported. Of course, in most practical cases, we wouldn't have the full search results to know this.

5.6. Observations. Those few experiments have confirmed some observations we had already made in a different context, and given new insights on the use of GE-SETS.

Compression. It is possible to achieve compression of the state space with GE-SETS, for typical SPIN applications. This is a non-trivial statement: GE-SETS are a compression technique which could very well result in a stored space larger than the number of bits required to record the plain data, according to Hamming.

Evolution of gains. Gains increase as state space get larger. In earlier work, not directly related to SPIN we had observed compression factor of over 100.

Time. There is a performance penalty in the use of GE-SETS. The run time cost appears also to decrease as state spaces become larger.

Partial Orders. Compression is achieved even in the case of a REDUCE'd search. The highest reported compression factor achieved here were actually realized in such a search. This would tend to show that the type of regularity exploited in GE-SETS is distinct from what "partial order" techniques exploit. This is worthwhile to note because a different behavior had been observed with ROBDDs [**V96**].

Sample Set. All our examples are "vintage PROMELA", namely processes communicating through asynchronous communication channels. The success we experience with such models bodes well for the possibilities of significant compression rates with GE-SETS in general.

5.7. Discussion. Where do GE-SETS fit in the SPIN verification philosophy? Their runtime cost, plus the added cost of generating an encoding do not make them a convenient replacement for either full state or bit state search modes. They would have to be viewed in a complementary way. The other search modes are the most appropriate to start debugging a specification, when the probability of finding an error is high and the duration of the runs low. GE-SETS-based search should come at the end, to try to achieve a "last", complete run, if it cannot be done in full state, and if we do have enough memory to do it, based on estimates of achievable compression and of the number of states (from bit state runs).

The early runs could also give critical information to define the clustering and the domain coding of GE-SETS layers. The automation of such a procedure is possible.

Finally, as we said earlier, we make no claim at this stage that all the algorithms available in SPIN in full state search can be supported as are. SPIN tends to store "marking" information with the state, which can be later modified *in situ*. Such a technique cannot be applied here, but rather would require the manipulation of a another GE-SETS when exploring loops. SPIN 's current philosophy is to mark data on the stack to avoid multiple copies of the state space, which was the case originally. We do not have such concern, since multiple copies introduce an opportunity for sharing. GE-SETS thus give another perspective on the classical verification problems, one for which SPIN's current solutions may not be adequate.

Comparison with related work

There is surprising little work done in the area of state space compression in SPIN. The key efforts we can relate to are Visser's adaptation of ROBDDs to SPIN [**V96**], which we have already mentioned, and SPIN's new (version 2.7.4.) COLLAPSE mode.

The COLLAPSE mode is a straightforward, but nevertheless quite elegant technique to compress the size of the state vector, without loss of information (unlike the BITSTATE mode). For each proctype, each reached state configuration is memorized and identified. What is stored in the global state vector is no longer the whole information, but the sequence of identifiers for each proctype configuration. Identifiers have to be a (compile-time) fixed multiple of 8 bits, from one to four.

The effect of collapsing on the GSM example are quite interesting. The amount of memory required was reduced from 140Mbytes (in 271s) to 38Mbytes (in 281s). The effective size of state vector was 14 bytes (down from 144). While the GE-SET-based results remain better, the compression rate achieved has dropped to about

7.8. The collapsing also has the advantage of being computed "on-the-fly", in true SPIN philosophy.

On PFTP, with loss and duplication, we have obtained the following results, with COLLAPSE3 (depth 58013):

	states	memory (Mb)	time (s)
SPIN	2315696	60.89	119.01
GE-SETS	2315696	6.81 (GE-SETS) + 3.06(SPIN)	516.05

In full state mode (depth 256099):

	states	memory (Mb)	time (s)
GE-SETS	22637670	6.68 (GE-SETS) + 11.88	4868.78

We were not able to get results with plain SPIN, even with COLLAPSE, because of lack of memory [7]. Notice that the amount of memory used by the GE-SET store is quite similar in REDUCE and NOREDUCE mode, in spite of a factor 10 of difference in the number of states. In this case, GE-SETS clearly exploit the same kind of regularity as the partial order technique.

One could wonder if it might be worthwhile to combine GETS and collapse compression. However, a quick computation shows that only half the storage space is used by the full state vector, while the other half is used to store the unique proctype configurations. We would have to code both to achieve any improvement over our previous results.

Visser also relates the application of a similar technique, developped independently, in combination with ROBDDs. The compression achieved is generally not as good as ours, and the measured runtimes are rather distressing.

6. Conclusion

We have presented a data structure optimised for a compact encoding of state spaces, and shown how well it performs in the context of SPIN. Our results show that there are major memory gains when a GE-SETS is used to store the state space, at the cost of a significant increase in execution time. The gains are not predictable *a priori*. The execution times tend to be linear in our observations.

The use of GE-SETS is also not entirely compatible with SPIN, in its current instance. They require tighter control over the real domains of variables, and the possibility of clustering variables together. On the algorithmic side, we haven't implemented all the algorithms supported by SPIN in full state search mode. While some could probably be easily adapted, it seems more appropriate to reengineer them to take advantage of the possibilities of GE-SETS, including the fact that replication is cheap in terms of storage space.

Acknowledgment: Our GE-SETS implementation was done by F. Gagnon, with improvements contributed over time by the author and J. Collette. Thanks to Prof. M. Ferguson for pointing out the Hamming argument about compression. Credit is due to D. Zampuniéris who gave us the initial inspiration to apply his Arbres Partagés to reachability analysis.

[7]The runtime image went beyond 140Mb, on a 128Mb machine.

References

[GGZ95] F. Gagnon, J-Ch. Grégoire, and D. Zampuniris, Sharing trees for 'on-the-fly' Verification, *Proc. FORTE'95*, 1995.

[H91] G. J. Holzmann, *Design and Validation of Computer Protocols*, Prentice Hall, Software Series, 1991.

[HP94] G. J. Holzmann and D. Peled, An improvement in formal verification, *Proc. FORTE'94*, 1994.

[V96] W. Visser, Memory efficient state storage in SPIN, *Proc. of the First* SPIN *Workshop*, 1995. `http://netlib.bell-labs.com/netlib/spin/ws95/spin95_abstracts.html`

[ZL95] D. Zampuniéris and B. Le Charlier, Efficient handling of large sets of tuples with sharing trees, *Proc. of the Data Compression Conference*, DCC95, March 1995.

INRS-TÉLÉCOMMUNICATIONS, 16, PL. DU COMMERCE, ILE DES SOEURS,, VERDUN, QC,, CANADA H3E 1H6

E-mail address: `gregoire@inrs-telecom.uquebec.ca`

DIMACS Series in Discrete Mathematics
and Theoretical Computer Science
Volume **32**, 1997

Not Checking for Closure under Stuttering

Gerard J. Holzmann and Orna Kupferman

ABSTRACT. The model checker SPIN works better with specifications that are *closed under stuttering*. Checking such specifications, SPIN can use its partial-order reductions. It is hard to check whether a given specification is closed under stuttering and it is pity to give up SPIN's partial-order reductions. We suggest an algorithm that, given a program P and a specification N of bad behaviors for P, checks the correctness of P with respect to an extension N' of N that is guaranteed to be closed under stuttering. In this check, SPIN can use its partial-order reductions. If P is correct with respect to N', we conclude that it is correct also with respect to N. If P is not correct with respect to N', we use the counter-example that SPIN provides to determine whether the program is correct with respect to N or that N is not closed under stuttering.

1. Introduction

To perform verifications with the model checker SPIN [**Hol91**], the user should provide two specifications. The first formalizes the behavior of the system to be verified; the second formalizes all possible violations of the correctness requirements. The first specification is usually given as a set of asynchronous processes, specified in the language of the verifier (PROMELA). During the verification itself, the interleaving product of these processes is computed as a *Büchi word automaton*. We refer to this product as the *program automaton* P. The second specification is given directly as a Büchi word automaton, in the syntax of PROMELA. We refer to this automaton as the *never-claim* N.

Verification is now reduced to checking the emptiness of the intersection of the automata P and N [**VW86**]. If the intersection is empty, it is guaranteed that no computation of the system violates the correctness requirement. If the intersection is not empty, SPIN generates an infinite word

1991 *Mathematics Subject Classification.* Primary 68Q65, 68Q22, 68M15; Secondary 68N20.

Key words and phrases. Verification, Partial Order Reduction, Stuttering, SPIN.

in the intersection, and presents it as a counter-example to the correctness requirement.

The intersection of P and N is computed *on-the-fly*, using a depth-first search algorithm. When a standard exhaustive depth-first search algorithm is used, it is irrelevant whether or not the language of N has special properties. For instance, it does not matter whether or not this language is *closed under stuttering*. There are, however, also optimized depth-first search algorithms that do make use of special properties that N might have. In particular, this is true for the *partial-order* reductions used in SPIN version 2. These reductions are guaranteed to preserve safety and liveness properties for an automaton N whose language is closed under stuttering [**HP94**].

Given a never-claim automaton N, a naive algorithm would test N for closure under stuttering and, according to the result, decide whether or not partial-order reductions can be applied. Testing N for closure under stuttering, however, is hard. Indeed, the problem of testing Büchi automata for closure under stuttering is PSPACE-complete [**PWW96**]. Ideally, we would like to use a method that avoids the test, but can still apply the partial order reduction algorithm in as many cases as possible. This paper presents such a method.

2. Preliminaries

Given an alphabet Σ, a *block over* Σ is a word in Σ^* all of whose letters are the same. We can view a word $w \in \Sigma^\omega$ as an infinite sequence of blocks and represent it as a function $w : \mathbb{N}^+ \to \Sigma \times \mathbb{N}^+$, such that the i'th block of a word w with $w(i) = \langle \sigma, n \rangle$ is σ^n. So, for example, the word 101100111000... is represented by $w(i) = \langle i \bmod 2, i \text{ div } 2 \rangle$. We use w_σ and w_n to denote w projected on its Σ and $\mathbb{N}^+$ elements, respectively; thus $w(i) = \langle w_\sigma(i), w_n(i) \rangle$. Note that we do not require that successive blocks are of different letters. A *normal form* for the word w is a function $w : \mathbb{N}^+ \to \Sigma \times \mathbb{N}^+$ where for every $i \in \mathbb{N}^+$ we have

$$w_\sigma(i) = w_\sigma(i+1) \to \text{ for all } j \geq i \text{ we have } w_\sigma(j) = w_\sigma(i) \text{ and } w_n(j) = 1.$$

That is, in a normal form, successive blocks are of different letters, except for the case where w has a suffix σ^ω. Then, the suffix is presented by infinitely many blocks of length 1. In the sequel, we assume that words are presented in this normal form.

We define an equivalent relation $S \subseteq \Sigma^\omega \times \Sigma^\omega$ as follows. For two words w and w', we have $S(w, w')$ iff $w_\sigma \equiv w'_\sigma$. Thus, two words are related by S if they agree on the labels of their blocks and may only disagree on the lengths of their blocks. If $S(w, w')$, we say that w' *stutters* w. We write $[w]$ to denote the set of all words w' for which $S(w, w')$. A language $\mathcal{L} \subseteq \Sigma^\omega$ is *closed under stuttering* iff for every word $w \in \Sigma^\omega$, we have that $w \in \mathcal{L}$ iff $[w] \subseteq \mathcal{L}$. Thus, if w is in $\mathcal{L}$, so are all words obtained from w by either *shrinking* or *stretching* each of its blocks.

For technical convenience we present our algorithm in terms of Büchi transition systems, rather than Büchi word automata. One can translate transition systems to word automata by moving labels from states to transitions. Similarly, by moving labels from transitions to states, one can translate word automata to transition systems.

A *Büchi transition system* $A = \langle \Sigma, W, R, W_0, L, F \rangle$ consists of an alphabet Σ, a set W of states, a total transition relation $R \subseteq W \times W$ (i.e., for every $w \in W$ there exists $w' \in W$ such that $R(w, w')$), a set W_0 of initial states, a labeling function $L : W \rightarrow \Sigma$, and an acceptance condition $F \subseteq W$. A *computation* of A is a sequence $\pi = w_0, w_1, w_2, \ldots$ of states such that for every $i \geq 0$ we have $R(w_i, w_{i+1})$. For a computation $\pi = w_0, w_1, w_2, \ldots$, let $Inf(\pi)$ denote the set of states that π visits infinitely often. That is,

$$Inf(\pi) = \{w \in W \ : \text{ for infinitely many } i \geq 0, \text{ we have } w_i = w\}.$$

We say that a computation is *accepting* iff $Inf(\pi) \cap F \neq \emptyset$.

For a system A, we use $\mathcal{L}(A)$ to denote the set of all words $\sigma_0 \cdot \sigma_1 \cdots \in \Sigma^\omega$ for which there exists an accepting computation $w_0, w_1, \ldots$ with $w_0 \in W_0$ and with $L(w_i) = \sigma_i$ for all $i \geq 0$. Thus, each transition system defines a subset of Σ^ω. We sometimes say that A accepts w, meaning that $w \in \mathcal{L}(A)$. We say that a transition system A is *empty* iff $\mathcal{L}(A) = \emptyset$; i.e., A has no accepting computation. We say that a transition system A is closed under stuttering iff $\mathcal{L}(A)$ is closed under stuttering.

3. Closing a Büchi Transition System under Stuttering

Given a Büchi transition system A, we construct a Büchi transition system A' such that $w' \in \mathcal{L}(A')$ iff there exists $w \in [w'] \cap \mathcal{L}(A)$. That is, every word accepted by A is also accepted by A'. In addition, A' also accepts all the words that stutter some word accepted by A. We call A' the *closure of A under stuttering*. The system A' should be more flexible than A in two senses. First, when A moves to a certain state, A' may prefer to stay in the current state, reflecting its tolerance towards "stretched" blocks. Second, when A proceeds to a certain state, A' may prefer jump to states that A reaches in several steps, reflecting its tolerance towards "shrunken" blocks. To preserve the acceptance condition we should further make sure that A' does not miss or add infinitely many visits to states in F.

Let $A = \langle \Sigma, W, R, W_0, L, F \rangle$. Before defining A', we need to define the *closure* of R and its *F-closure*.

- For two states w and w', we have that $cl(R)(w, w')$ holds iff there exists $n \geq 1$ and $w_1, w_2, \ldots, w_n$ such that $w_1 = w$, $w_n = w'$, and for all $1 \leq i \leq n-1$ we have $R(w_i, w_{i+1})$ and $L(w_i) = L(w_1)$.
- For two states w and w', we have that $cl_F(R)(w, w')$ holds iff there exists $n \geq 1$, $1 \leq j \leq n$, and $w_1, w_2, \ldots, w_n$ such that $w_1 = w$, $w_j \in F$, $w_n = w'$, and for all $1 \leq i \leq n-1$ we have $R(w_i, w_{i+1})$ and $L(w_i) = L(w_1)$.

Intuitively, $cl(R)$ relates two states w and w' iff there exists a path from w to w' all of whose states are labeled as w. The relation $cl_F(R)$ is the same, only that at least one accepting state is visited in the path from w to w'.

Given A, let $A' = \langle \Sigma, W \times \{0,1\}, R', W_0 \times \{0\}, L', W \times \{1\} \rangle$, where for every $w \in W$ we have that $L'(\langle w,0 \rangle) = L'(\langle w,1 \rangle) = L(w)$ and $R'(\langle w,c \rangle, \langle w',c' \rangle)$ holds iff one of the following holds.

1. $w = w'$ and $c = c' = 0$.
2. $cl(R)(w,w')$ and $c' = 0$.
3. $cl_F(R)(w,w')$ and $c' = 1$.

That is, the system A' handles tolerance towards stretched blocks by self loops, and it handles tolerance towards shrunken blocks by proceeding according to the closure. The distinction between $cl(R)$ and $cl_F(R)$ as well as the duplication of the states, serves two goals. First, A' can loop only in states that are not accepting. Otherwise, it might have accepted also words with suffixes σ^ω that stutter no word accepted by A. Second, when A' reads a block of length 1 whose corresponding block in a word accepted by A visits F, it can visit an accepting state and "keep tracking" A in its next block.

THEOREM 3.1. *For every Büchi transition system A and its closure under stuttering A', the following hold.*

(1): $\mathcal{L}(A) \subseteq \mathcal{L}(A')$.
(2): *A' is closed under stuttering.*
(3): *A is closed under stuttering iff $\mathcal{L}(A) = \mathcal{L}(A')$.*

Note that the number of reachable states of A' is at most twice that of A. Recall that SPIN gets its input as a word automaton (rather than a transition system). In this case, the construction of N' is slightly more complicated. We may have to add up to $|W| \times |\Sigma|$ new states (one state for each transition in N). The following section shows how, in cases where the size of N' remains close to that of N, SPIN can use N' (with partial-order reduction), in order to check N, without first checking N for closure under stuttering.

4. The Algorithm

Given P and N, let N' be the closure of N under stuttering. Our algorithm is summarized in the figure below.

Instead of using N, SPIN uses N' and computes its intersection with P. Since the automaton N' is known to be closed under stuttering, this can be done using partial-order reductions. If the intersection is empty then, as $\mathcal{L}(N) \subseteq \mathcal{L}(N')$, it is guaranteed that the intersection of N and P is empty as well. Hence, P is proven to be correct with respect to N.

If the intersection of N' and P is not empty, SPIN generates a counterexample x accepted by both N' and P. SPIN now checks whether x is accepted by N. Since x is a single word, this can be done easily by projecting the computations of N on x. If x is accepted by N then, as it is also accepted

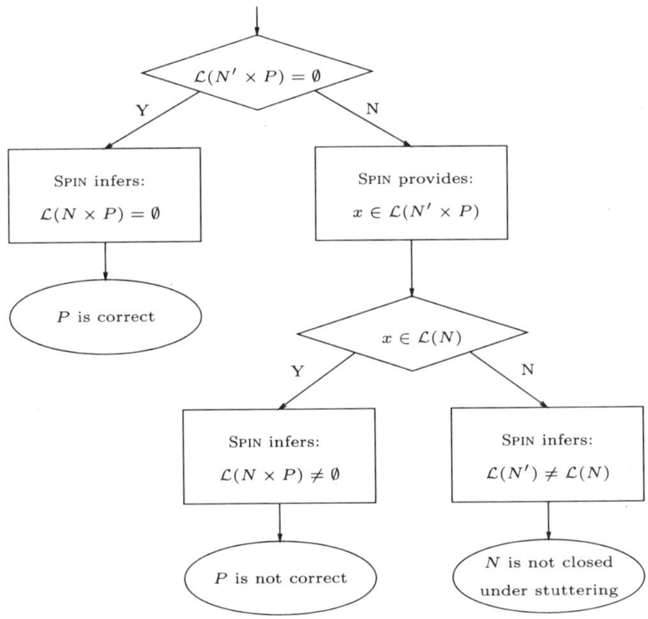

by P, we found a violation of the correctness, and P is proven to be incorrect with respect to N. If x is not accepted by N then, as it is accepted by N', we proved that N is not closed under stuttering. The user can then choose to repeat the verification with a revised claim, that is closed under stuttering, or to check the original N without partial-order reduction.

Note that in some cases, the proof of P with respect to N can proceed successfully using N' and the optimized partial-order reduction algorithm, even when the original N is not closed under stuttering. To check merely the closure of N under stuttering, our algorithm would be neither efficient nor helpful. First, P is usually much bigger than N. Second, when P is proved to be either correct or incorrect, no information on being N closed under stuttering is obtained. The question we want to answer, however, is not whether N is closed under stuttering, but whether P is correct with respect to N, and we want to do so as efficiently as possible.

5. An Example

The figure below shows a Büchi transition system N and its closure N' under stuttering. On the bottom are three sample programs P_1, P_2, and P_3.

We demonstrate the execution of our algorithm with respect to these programs. The intersection with N' is empty for program P_1, proving P_1's correctness with respect to N without further checks. For the other two programs, the intersection with N' is not empty. Our algorithm now checks the membership of a counter-example (which in this case equals the original program) in N. For P_2, membership in N holds, proving that N is not closed under stuttering. For P_3, membership does not hold, proving that P is not

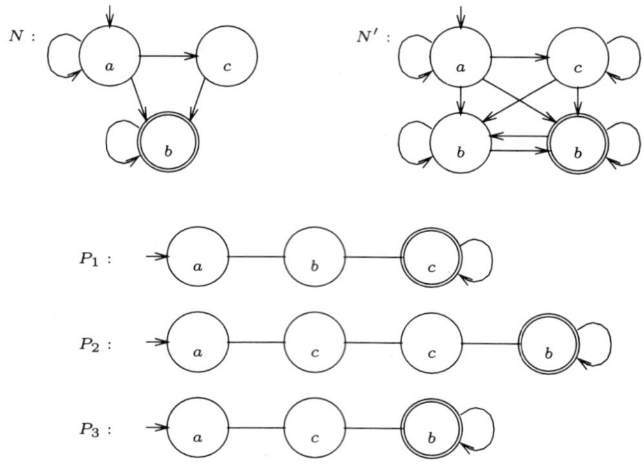

correct with respect to N. In the case of both P_1 and P_3 we can establish the program's (in)correctness, using N' with partial-order reduction, despite the fact that N itself is not closed under stuttering. In the case of P_2, we obtain an answer about the closure of N under stuttering.

References

[Hol91] G.J. Holzmann, *Design and Validation of Computer Protocols.* Prentice Hall, Software Series, 1991.

[HP94] G.J. Holzmann, D. Peled, An Improvement in Formal Verification, *Proc. 7th Int. Conf. on Formal Description Techniques*, Berne, Switzerland, 1994, 177–194.

[PWW96] D. Peled, T. Wilke, and P. Wolper. An Algorithmic Approach for Checking Closure Properties of ω-regular Languages. *Proc. 7th Conf. on Concurrency Theory*, Pisa, Italy, 1996.

[VW86] M.Y. Vardi and P. Wolper. An Automata-Theoretic Approach to Automatic Program Verification. *Proc. of the First Symposium on Logic in Computer Science*, 322–331, Cambridge, June 1986.

Bell Laboratories 2C-521, 700 Mountain Avenue, Murray Hill, NJ 07974
E-mail address: {gerard,ok}research.bell-labs.com

DIMACS Series in Discrete Mathematics
and Theoretical Computer Science
Volume **32**, 1997

On Nested Depth First Search

Gerard J. Holzmann, Doron Peled, and Mihalis Yannakakis

ABSTRACT. We show in this paper that the algorithm for solving the model checking problem with a nested depth-first search can interfere with algorithms that support partial order reduction. We introduce a revised version of the algorithm that guarantees compatibility. The change also improves the performance of the nested depth-first search algorithm when partial order reduction is not used.

1. Introduction

The model checking problem for concurrent systems can be solved elegantly by modeling the system and its correctness requirements as automata on infinite words (i.e., Büchi Automata), and by verifying that the language of the product of system and requirement automata is empty. The emptiness problem itself would normally require the computation of the strongly connected components of the product system. This computation can be avoided with a nested depth-first search procedure, as discussed in [**CVWY92, GodHol93, Hol92**]. The memory requirements of a reachability analysis, based on either depth-first or breadth-first search order, can be reduced further with the help of property preserving reduction strategies. One such method was described in [**Peled94, HP94**]. We recently discovered that the two methods are not completely compatible, which means that a direct combination will result in an incomplete model checking procedure.

Section 2 first gives a synopsis of the basic and nested depth-first search methods. Section 3 discusses reduction, and illustrates the undesirable side-effects of a direct combination of the algorithms. Section 4 describes a modification of the nested depth-first search algorithm that avoids the problem. In Section 5, finally, we discuss a variation of the algorithm from [**Hol92**] that can be used for proving absence of starvation, but that does not appear

---

1991 *Mathematics Subject Classification.* Primary 68Q65, 68Q22, 68M15; Secondary 68N20.

Key words and phrases. Verification, Depth-First Search, Partial Order Reduction, SPIN.

```
proc dfs(s)
    if error(s) then report error fi
    add s to Statespace
    for each successor t of s do
      if t not in Statespace then dfs(t) fi
    od
end
```

FIGURE 1. Basic Depth-First Search Algorithm, Used For Reachability Analysis

to be compatible with the reduction strategy. An alternative technique, that is compatible, is also discussed.

2. Basic Depth-First Search

A basic depth-first search algorithm generates and examines every global state that is reachable from a given initial state ι, as illustrated in Figure 1.

After a first check for the validity of the state and its properties, the state descriptor is entered into a global Statespace, usually with the help of standard hash-table lookup procedures. A recursive call to the search procedure is then made for each state that is reachable from this state in one atomic execution step, i.e., for each possible successor in the reachability graph that is not already represented in Statespace.

Note that Statespace needs to represent only the nodes of the reachability graph; the representation of the edges between the nodes is not needed. Note also that no information is needed about the presence of any state on the depth-first search stack.

3. Nested Depth-First Search

The procedure from Figure 1 cannot be used directly to solve a model checking problem, but it can readily be adapted. In model checking, the reachability graph, partially represented in the Statespace structure, is used to define a Büchi automaton. The acceptance conditions in this automaton are usually derived from a special property automaton that is added to the global system, as described in [**VW86, Hol92**].

The model checking problem in this context reduces to the problem of detecting the existence in the graph of accepting states that are part of a strongly connected component and therefore reachable from themselves.

The procedure in [**CVWY92**] applies a nested depth-first procedure to solve this problem, as illustrated in Figure 2. When the normal search retracts to an accepting state, a second search is started, to search for a cycle through this state. If the second search fails to find a cycle, the first search

```
proc dfs(s)
    if error(s) then report error fi
    add {s,0} to Statespace
    for each successor t of s do
      if {t,0} not in Statespace then dfs(t) fi
    od
    if accepting(s) then seed:=s; ndfs(s) fi
end
proc ndfs(s) /* the nested search */
    add {s,1} to Statespace
    for each successor t of s do
      if {t,1} not in Statespace then ndfs(t) fi
      else if t==seed then report cycle fi
    od
end
```

FIGURE 2. Nested Depth-First Search

resumes from the point where it was interrupted. It is shown in [**GodHol93**] that adding two bits to every global state that is stored in the Statespace that is constructed by the algorithm from Figure 1 suffices to separate the two searches. That is: the size of Statespace need not double as initially suggested in [**CVWY92**], but can remain virtually unchanged. The search time, however, can double when all states are reachable in both the first and the second search, and no cycles through accepting states exist.

Notice that also in the modified algorithm, there is still no need to store the edges of the graph, nor to access the depth-first search stack.

4. Reduced Search

The statespace that is constructed with the depth-first search procedure can be reduced substantially if we limit the number of successor states that is explored from each reachable state.

In [**Peled94**] a reduction of this type is discussed that preserves the Büchi acceptance properties from the full statespace also in the reduced statespace, provided that three conditions are satisfied. Two of the three conditions deal with dependency between execution steps, and the visibility of individual execution steps. They do not alter the properties of either the basic or the nested the depth-first search.

A third condition, however, introduces a dependency on the information that is stored in the depth-first search stack. The condition states that a reduction of the set of successor states from state s is invalid if at least one of those successor states appears on the depth-first search stack. This means

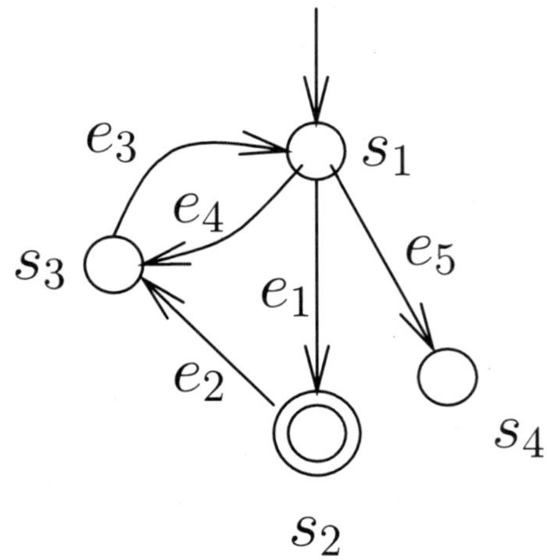

FIGURE 3. Failure of the Reduced Nested Depth-First Search

that reachability properties during a *reduced* search that is based on this condition can depend on the precise search order and the contents of the search stack at each point during the search.

A direct combination of the partial order reduction technique presented in [**Peled94, HP94**] with the nested depth-first search method described in [**CVWY92**] therefore results in an incorrect algorithm. Since the contents of the depth-first search stacks in the first and in the second search in the nested depth-first search algorithm from Figure 2 differ, the state space that is constructed during the first and the second part of the search need no longer be equal during a reduced search. This means that the second search may be incapable of proving that an accepting state is reachable from itself, even if it is part of a strongly connected component in the reachability graph that is constructed during the first search.

The example in Figure 3 illustrates the problem. Consider the case where the first search reaches state s_1. Its successors are s_2 and s_3. In the full state space, there is another successor s_4, but it is not selected. First edge e_1 is taken, leading to accepting state s_2. Now, the first search continues to search all the nodes accessible from s_2. When backtracking to s_2, the second search starts, with s_2 as its seed state, looking for a cycle. Such a cycle exists by taking the edges e_2, e_3 and then e_1 again. However, once reaching s_1 during the second search, a cycle consisting of e_3 and e_4 is found. Notice that this cycle was not found in the first search, as s_3 was not on the search stack of the first search while searching for successors of s_1. Closing this cycle with the edge e_4 in the second search makes the second search select an alternative subset of successors, e.g., s_4, via the edge e_5. But now, it is possible that the cycle through seed state s_2 remains undetected.

5. The Correction

To correct the algorithm it will suffice if we can guarantee that the second depth-first search always explores the same states that are found in the first depth-first search. We will present two changes to accomplish this.

5.1. Intersection with Search Stack from First Search. The problem in Figure 3 occurs when the second depth-first search reaches a state that exists also on the first search stack (state s_1). Continuing the second search from this state, the search can now reach states with a different (longer) search stack in the second depth-first search than in the first. This means that the third condition from the partial order reduction algorithm is applied to different states, and the second search might not follow the first search. Where the sets of successor states that is selected in the two phases of the search can differ.

The solution to this problem is remarkably simple: when the second depth-first search reaches a state that exists also on the stack from the first search, the search can be *terminated*. The reason is that reaching such a state from the seed state implies immediately that a path exists that leads back to the seed state: it is the path of states on the first depth-first search stack that starts at the point of intersection. This property is independent of the use of the partial order reduction. Stopping the search when reaching a state on the first search stack shortens the search, it improves the performance of the algorithm, and it allows for shorter counter-examples to an invalid correctness claim to be generated.

The change in the algorithm of Figure 2 from [**CVWY92**] is shown in Figure 4. In the new version of this algorithm, there is no longer a need to store the seed state, but we do need information about the presence of states on the search stack.

5.2. Preserving Information Between Searches. The first correction of the nested depth-first search algorithm improves its performance and it eliminates the error that occurred in Figure 3. But, it does not completely solve the problem.

A failure can still happen. First note that the first depth-first search can backtrack from a graph component without performing a second search. (The second search is only initiated from accepting states.) While searching another component of the graph, that does contain accepting states, there can be a transition that leads back to the earlier states, that do not as yet appear in the second state space. This return transition would not be followed in the first search, since it leads only to previously visited states. During a second search, however, the transition must be followed, A similar scenario now exists, in which the states are visited with a different search stack in each of the two searches, and the reachability properties are not preserved.

To eliminate also this problem, one must preserve information about the selection of successors between the two searches. There are several possibilities:

- Add an integer that holds a "selection number." The partial order reduction algorithm can make several attempts to select a safe (reduced) subset of successor states. In SPIN, these selections correspond to process numbers [**HP94**]. When all attempts to chose a safe selection fail, a complete expansion of all successors is done. The first depth-first search would keep the selection number for the use of the second depth-first search. The second depth-first search would generate the sets of successors (*ample sets*, in the terminology of [**Peled94**]) according to the same order, and use the selection number that was held by the first depth-first search, ignoring its own cycle closing. A selection number of zero would mean selecting all successors from the current state.
- A more modest solution uses only one bit as a selection number. In this case, if a selection caused closing a cycle in the first depth-first search, no alternative subset is sought, and the entire set of successors is generated from the current node. The selection bit communicates to the second depth-first search whether to use its first subset or to do a full successor expansion.

The addition of the selection bit is a small modification of the reduction algorithm discussed in [**HP94**] that we will not elaborate further here.

6. Absence of Starvation

In [**Hol92**] a different version of a nested depth-first search is described to solve a slightly different type of problem. The problem here is the detection of cycles in the reachability graph that do *not* contain any user-defined *progress* states. Any infinite execution that contains only finitely many traversals of progress states corresponds to starvation.

The algorithm from [**Hol92**] is shown in Figure 5 (shown here in the same format as Figures 1 and 2). The differences with Figure 2 are as follows. From every reachable state we start the nested search, but in the second search no traversals through progress states are allowed. Whenever a cycle is closed on the stack from this second depth-first search, an error can be reported. In this variant of the search, therefore, also the non-reduced variant relies on information from the search stack. (Note that an explicit search stack must be maintained here only for the *second* search, while in Figure 4 the explicit stack must be maintained only for the *first* search.)

A direct combination of this algorithm with the reduction strategy from [**Peled94**] fails for the same reasons as before, but this time the correction also fails. Note that when the second search intersects the search stack from the first depth-first search, there does not necessarily exist a valid path back

```
proc dfs(s)
    if error(s) then report error fi
    add {s,0} to Statespace
    add s to Stack
    for each (selected) successor t of s do
      if {t,0} not in Statespace then dfs(t) fi
    od
    if accepting(s) then ndfs(s) fi
    delete s from Stack
end
proc ndfs(s) /* the nested search */
    add {s,1} to Statespace
    for each (selected) successor t of s do
      if {t,1} not in Statespace then ndfs(t) fi
      else if t in Stack then report cycle fi
    od
end
```

FIGURE 4. Optimized Nested Depth First Search, Compatible with Reduction

```
proc dfs(s)
    if error(s) then report error fi
    add {s,0} to Statespace
    for each successor t of s do
      if {t,0} not in Statespace then dfs(t) fi
    od
    ndfs(s) /* different */
end
proc ndfs(s) /* the nested search */
    if s is Progress State then return fi /* new */
    add {s,1} to Statespace
    add s to Stack /* new */
    for each successor t of s do
      if {t,1} not in Statespace then ndfs(t) fi
      else if t is in Stack then report cycle fi /* different */
    od
    delete s from Stack /* new */
end
```

FIGURE 5. Nested Depth First Search for Non-Progress Cycles

```
never { /* non-progress: ◇□¬progress */
    do
    :: skip
    :: !progress -> break
    od;
accept: do
    :: !progress
    od
}
```

FIGURE 6. Non-Progress Büchi Automaton, expressed as Never Claim

to the root of the second depth-first search (the seed state). (The path that exists is invalid if it contains progress states.)

Although we do not know how to modify the algorithm from Figure 5 for compatibility with partial order reduction, there is a simple alternative. The non-progress property can trivially be expressed as a model checking problem, for instance by expressing it with the LTL property: $\Diamond\Box\neg progress$, where $progress$ is a boolean function that yields $true$ in progress states, and $false$ otherwise. This LTL property can be translated into the Büchi automaton shown in Figure 6, which can be used to solve the problem with the algorithm from Figure 4.

The method from Figures 4 and Figure 6 are part of the latest version of the model checker SPIN (Version 2.9). The automaton from Figure 6 does not have to be specified by the user, but is automatically generated by SPIN when the user selects a search for non-progress cycles. The addition of the two-state non-progress automaton in Figure 6 can in the worst case double the memory requirements of the search based on Figure 4, compared to a direct application of the algorithm from Figure 5 (avoiding the need for the automaton from Figure 6). The algorithm from Figure 4, however, can be optimized significantly by combining it with a partial order reduction strategy. The measurements in [**HP94**] suggest that a reduced search based on Figure 4 should in almost all cases outperform an exhaustive search based on Figure 5.

7. Conclusion

The original algorithms for performing nested depth first searches to detect the presence of acceptance cycles ([**CVWY92**]) or the presence of non-progress cycles ([**Hol92**]) are both incompatible with partial order reduction, such as proposed in [**Peled94, HP94**]. We have discussed a modification of one of the two algorithms that secures compatibility with reduction methods, and that suffices to solve both search problems efficiently and uniformly.

Acknowledgement: The authors are most grateful to Michael Ferguson, whose keen observations lead us to the discovery of the incompatibility of the nested depth first search algorithms with partial order reduction.

References

[CVWY92] C. Courcoubetis, M. Vardi, P. Wolper, M, Yannakakis, Memory-efficient algorithms for the verification of temporal properties, Formal methods in system design 1 (1992) 275–288.

[GodHol93] P. Godefroid, G.J. Holzmann, On the Verification of Temporal Properties In Proc. PSTV93,, Protocol Specification Testing and Verification, Liege, Belgium, 1993, North-Holland, 109–124.

[Hol92] G. J. Holzmann, *Design and Validation of Computer Protocols*, Prentice Hall Software Series, 1992.

[HP94] G.J. Holzmann, D. Peled, An improvement in formal verification, *7th International Conference on Formal Description Techniques*, Berne, Switzerland, 1994, 177–194.

[Peled94] D. Peled. Combining partial order reductions with on-the-fly model checking. *Formal Methods in System Design* 8 (1996), 39–64.

[VW86] M.Y. Vardi, P. Wolper, An automata-theoretic approach to automatic program verification, *1st Annual IEEE Symposium on Logic in Computer Science*, 1986, Cambridge, England, 322–331.

Bell Laboratories 2C-521, 700 Mountain Avenue, Murray Hill, NJ 07974
E-mail address: {gerard,doron,mihalis}research.bell-labs.com

DIMACS Series in Discrete Mathematics
and Theoretical Computer Science
Volume **32**, 1997

Modelling and Analysis of a Collision Avoidance Protocol Using SPIN and UPPAAL

Henrik Ejersbo Jensen, Kim G. Larsen, and Arne Skou

ABSTRACT. This paper compares the tools SPIN and UPPAAL by modelling and verifying a Collision Avoidance Protocol for an Ethernet–like medium. We find that SPIN is well suited for modelling the untimed aspects of the protocol processes and for expressing the relevant (untimed) properties. However, the modelling of the media becomes ackward due to the lack of broadcast communication in the PROMELA language. On the other hand we find it easy to model the timed aspects using the UPPAAL tool. Especially, the notion of *committed locations* supports the modelling of broadcast communication. However, the property language of UPPAAL lacks some expressivity for verification of bounded liveness properties, and we indicate how timed testing automata may be constructed for such properties, inspired by the (untimed) checking automata of SPIN.

1. Motivation

During the last few years, the SPIN tool [**Hol91**] has attracted much interest from university people teaching formal methods and from industrial developers. Its merits include a simple yet powerful design language based on asynchronous channels, as well as an expressive logic for property verification. However, until recently [**TC96**], it has not been possible to apply SPIN to verification of real time systems.

In this paper we compare the existing (untimed) SPIN with a recent real time tool, UPPAAL, which is based on timed automata specifications. This is done through application of the tools on a small protocol[1] for collision avoidance on an Ethernet-like broadcast medium. The untimed properties are verified on a PROMELA model and the real time properties are verified on a timed automata model.

1.1. The Example.

We assume that a number of stations are connected on an Ethernet-like medium, see Figure 1, that is, the basic protocol is of type CSMA/CD.

1991 *Mathematics Subject Classification.* Primary 68Q60; Secondary 68Q22.

The authors were supported in part by BRICS, **B**asic **R**esearch in **C**omputer **S**cience, Centre of the Danish National Research Foundation.

[1]Our example is inspired by a recent paper by Karsisto and Valmari [**KV96**]

On top of this basic protocol, we want to design a protocol without collisions[2], that is, we want to guarantee a lower bound on the transmission delay of a buffer - assuming that the medium does not loose or corrupt data and also assuming that the stations function properly[3]. The basic (obvious) idea of the protocol is to introduce a dedicated master station, which in turn asks the other stations if they want to transmit data to another station. However, the master has to take into account the possible buffer delays within the receiving stations. Hence, we want the protocol to enjoy the following properties:

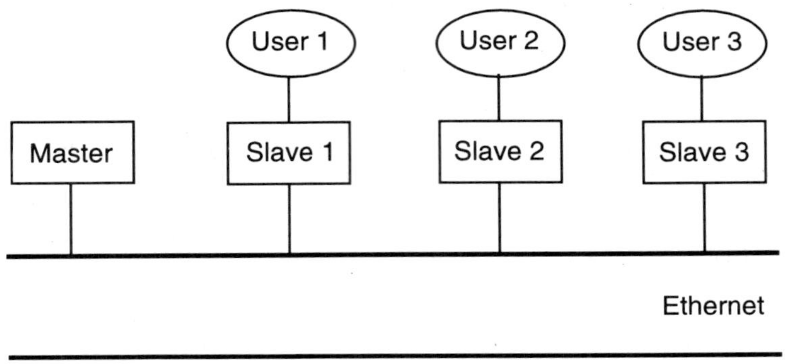

Figure 1: The Ethernet

- Collision cannot occur.
- Data which are received, have been transmitted by a sender.
- Assuming error-free transmission, there is a known upper bound on the transmission delay.

Assuming that we know the buffer delays introduced by the medium and the slave stations, it should not be difficult to make a reasonable estimate of the upper bound by hand — assuming that the master makes enquiries according to a round-robin strategy. However, if we want to exploit the potential parallelism, it might very well be that one can find a more intrinsic strategy, which decreases the upper bound. Hence we would like to check for the following additional property:

- Does there exist a slave schedule with an upper bound being smaller than the sum of the individual slave delays?

1.2. Paper Organization. In section 2 we present a PROMELA model for the protocol, and we verify the above untimed properties. In section 3 we introduce UPPAAL and present the protocol model — applying the notion of committed locations to model broadcast. We verify the remaining timed properties and we indicate how one may extend UPPAAL with a more expressive logic than the existing one. Finally, in section 4, we sum up and compare the two tools on different aspects.

2. Modelling and Verification in SPIN

In this section we describe the modelling and verification of the Collision Avoidance Protocol using the design language PROMELA [**Hol91**] and the validation tool

---

[2]Applicable for e.g. real time plants.

[3]Well known classical protocols exist to handle these error cases.

SPIN [**Hol91, Hol**]. We first present the PROMELA design and we then discuss the verification of this. The interesting verifications include verifying that no collisions occur and that data is delivered correctly between users.

2.1. The PROMELA design. Our basic modelling of the protocol consists of four slave processes and one master process. Master and slaves communicate using an independent process which models the communication medium (the Ethernet). The need for an independent process to model the medium is influenced by the need to model the mediums broadcast behaviour and the fact that we want the medium to be able to 'detect' collisions. The medium is modelled to be erroneous in the sense that it can loose messages, and to model the fact that slaves can loose messages independently, loss is modelled explicitly in the slaves. Finally the above model exists in a 'testing environment' consisting of one user process per slave. Slaves ask their user processes for data when they are enquired from the master, and in response user processes can either send data or indicate that they are not interested in sending anything. In this case the enquiry moves on to the next slave in round–robin fashion.

Message Format. All inter process communications are modelled by message passing on channels. Four different types of channels are used, each of zero capacity implying rendez-vous communication, see Figure 2. As we have used an independent process to model the medium, it is quite natural to model the communication between the master and the medium and between the slaves and the medium as synchronous. Messages from master and slaves to the medium are communicated on the channel `to_medium`, and messages from the medium are broadcasted using the channels `from_medium[N]` where `N` indicates the number of slaves plus one (the master). Messages between slaves and their users are communicated on channels `in[N]` and `out[N]`.

All messages passing the medium are formatted using three fields indicating the sender, the receiver and the type of data sent. The master's id is 0 and the slaves are numbered from 1 to `N`−1. Data types include `ENQ` indicating an enquiry from the master, and numbers in the range 0 to `N`−1 modelling data sent between users.

```
chan to_medium = [0] of {byte, byte, byte}
 /* channel to the medium : {sender, receiver, data} */

chan from_medium[N] = [0] of {byte, byte, byte}
/* channels from medium to each of master and slaves */

chan in[N] = [0] of {byte, byte}
/* channels between user and slave parts */

chan out[N] = [0] of {byte}
/* channels between user and slave parts */
```

FIGURE 2. Message Channel Formats

The Master. The master process, see Figure 3, passes enquiries round to slaves in round–robin fashion. Having sent an enquiry the master waits until it receives data and then it sends an enquiry to the next slave. If messages are lost the master will wait until 'silence' is detected and then it will send the next enquiry.

Detecting silence is modelled by the `timeout` statement of SPIN which by default blocks until nothing in the full system is executable and then it becomes executable itself. Actually the master always wait for `timeout` before sending the next enquiry. This is because the silence of the system either is due to message loss as described above, or that a message has been correctly delivered at all slaves and at the master. In the later situation all processes will be listening for data i.e. no statements are executable besides `timeout`.

```
proctype master()
{
        byte sender;
        byte receiver;
        byte data;
        byte next;

        next=1;
        to_medium!0,next,ENQ;      /* enquiring the first slave */

        do
        :: from_medium[0]?sender,receiver,data
        :: timeout -> next=next%(N-1)+1;to_medium!0,next,ENQ
        od
}
```

FIGURE 3. The Master

The Medium. As mentioned we use an independent process to model the behaviour of the communication medium; the Ethernet. The fundamental property of the medium is that it broadcasts messages to all processes connected to the medium. In PROMELA there exists no broadcast primitive, so we model this explicitly using an approach where the medium, when receiving data from either the master or a slave, sends this data to all other processes connected to the medium in turn. Of course we would like this sequence of events to be atomic, but although PROMELA features an `atomic` primitive to ensure atomicity on a sequence of statements, this can not be used in our situation, as in general no blocking statements are allowed in `atomic` enclosings. The atomicity requirement says that no processes connected to the medium may start sending before the broadcast sequence of the medium is done. We model this by not allowing the medium to listen before the end of the broadcast. However, communications between users and slaves are allowed to interleave the broadcast.

The medium process itself can not loose messages (ignore them). This is because we want the possibility of the individual slaves to loose messages and therefore loss is modelled explicitly in the slaves.

The part of the `medium` code at the `accept_collision` label, see Figure 4, is for verification and will be discussed further in section 2.2.

The Slaves. The slave processes listens for messages at the medium, and whenever a message is communicated the slaves either ignores it (losses) or reads it. In the later case the slaves now determine what type of message is received and whether this message is addressed to them or not. If the message contains data (not enquiry) addressed to the slave, the slave sends the data along to its user. If the message is an enquiry from the master, the enquired slave asks its user whether

```
proctype medium()
{
        byte sender;
        byte receiver;
        byte data;
        byte i;

        do
        :: to_medium?sender,receiver,data ->

           i=0;
           if
           :: (1) -> do
                     :: i<=N-1 ->
                          if
                          :: i==sender -> skip
                          :: i!=sender -> from_medium[i]!sender,receiver,data
                          fi;
                          i=i+1
                     :: i>N-1 -> break
                     od
           :: to_medium?sender,receiver,data -> goto accept_collision
           fi
        od;

accept_collision: do
                  :: (1) -> skip
                  od                    /* collision detected */
}
```

FIGURE 4. The Medium

it is interested in sending messages or not. In the former case the slave passes the messages on to the medium. See Figure 5.

The Users. The final part of the PROMELA design consists of the environment of the protocol, namely the user processes communicating with the slaves. The users can be seen as a sort of an testing environment for verifying the protocol, and further commenting will be given in section 2.2.

During the development of the above design we used the simulation facility of the validation tool SPIN to support early fault detection. Especially the message sequence chart (MSC) facility was very supportive in examining the communication pattern of our design.

2.2. Verification in SPIN. Using the simulation facilities of SPIN gives a first confidence in the correct behaviour of the design, as we can 'run' the protocol and see that there exists behaviours of our model that meet our expectations. Running the simulations we can for instance see, that there is a possibility that messages get lost and that the master in these situations sends on the enquiry to the next slave. Fortunately, we can also see that messages (data) can be sent correctly from user to user.

The main aim of the designed protocol is to avoid collisions in the medium, so we want to verify this property. Furthermore, we want to verify that when data is sent between users – and not lost – then receiving users will actually get the

```
proctype slave(byte id)
{
        byte sender;
        byte receiver;
        byte data;
        byte ny_receiver;
        byte ny_data;

        do
        :: from_medium[id]?sender,receiver,data ->
           if
           :: data!=ENQ ->
                if
                :: receiver==id -> out[id]!data
                :: receiver!=id -> skip
                fi
           :: data==ENQ && receiver==id ->
                in[id]?ny_receiver,ny_data;
                to_medium!id,ny_receiver,ny_data
           :: data==ENQ && receiver!=id -> skip
           fi
        :: from_medium[id]?sender,receiver,data         /* message loss */
        od
}
```

FIGURE 5. The Slaves

```
proctype user(byte id; chan cin,cout)
{
        byte data;

        do
        :: cout?data ->
           if
           :: data!=id -> goto accept_wrong_data
           :: data==id -> skip
           fi
        :: cin!0,0                              /* not interested */
        :: cin!(id%(N-1)+1),(id%(N-1)+1)
        od;

accept_wrong_data:

        do
        :: (1) -> skip                          /* wrong data received */
        od
}
```

FIGURE 6. The Users

data that is addressed to them. That is, messages can not get miss–directed in the medium.

Verifying that collisions are avoided is done by forcing the `medium` to enter an acceptance cycle if collisions occur, see Figure 4. The `medium` process will enter the acceptance cycle labelled `accept_collision` if `medium` can participate in two consecutive synchronizations on channel `to_medium`. By consecutive we mean with no broadcast delivery in between the receivings. Using SPIN to perform a full state space search including partial order reductions verifies that no acceptance cycles exist in the design.

Verifying that messages are correctly delivered to user processes is done using the testing environment consisting of the `user` processes. On request from their slaves users either indicate that they are not interested in sending data, or they send along data to their 'successor' which is simply interpreted as the user with process id one greater than the sending user. To guarantee that a correct unique receiving can be verified, the data send to the successor will be the id of the successor. Unique user id's are passed to user and slave processes on instantiation. Now, the `user` processes are forced into an acceptance cycle if they receive data from their slaves that do not correspond to the user id. See Figure 6. Using SPIN we verify that no acceptance cycles of the above nature exists.

3. Modelling and Verification in UPPAAL

The main purpose of the Collision Avoidance Protocol is to ensure an upper bound on the user communication delay in the Ethernet by avoiding collisions. It is therefore interesting to include *timing* in the design and to verify timing properties of the protocol.

In this section we describe the modelling and verification of the Collision Avoidance Protocol including *real–time* using the verification tool UPPAAL. We consider the process of transforming our PROMELA design to UPPAAL format. First, we introduce the UPPAAL tool and its underlying model. Then, we consider the design of the timed protocol and finally we consider the verification of timing properties.

3.1. The UPPAAL tool. UPPAAL is a tool suite for automatic verification of safety and bounded liveness properties of real-time systems modeled as networks of timed automata extended with data variables [**YPD94, LPY95a, BLL⁺95**], developed during the past two years. In this section, we summarize the main features of UPPAAL, applications to various case–studies and provide pointer to the theoretical foundation.

UPPAAL consists of a graphical user interface based on Autograph, that allows system descriptions to be defined graphically and a model-checker that combines *on-the-fly* verification with a *symbolic* technique reducing the verification problem to that of solving simple *constraint systems* [**YPD94, LPY95a**]. The current version of UPPAAL is able to check for invariant and reachability properties, in particular whether certain combinations of control-nodes of timed automata and constraints on variables are reachable from an initial configuration. Bounded liveness properties can be checked by reasoning about the system in the context of a testing automata. In order to facilitate debugging, the model-checker will report a *diagnostic trace* in case the verification procedure terminates with a negative answer [**LPY95b**].

The current version of UPPAAL is implemented in C++. An overview of UPPAAL is shown in Figure 7, and contains the following:

`atg2ta`: A compiler from the graphical representation (`.atg`) of a network of timed automata, to the textual representation in UPPAAL (`.ta`).

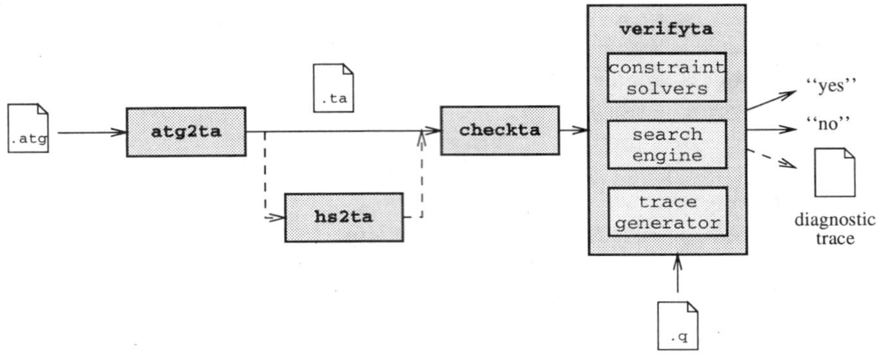

FIGURE 7. Overview of UPPAAL

`hs2ta`: A filter that automatically transforms linear hybrid automata where the speed of clocks is given by an interval into timed automata [**OSY94**], thus extending the class of systems that can be analyzed by UPPAAL.

`checkta`: Given a textual representation (in the `.ta`-format) of a network of timed automata, `checkta` performs a number of simple but in practice useful syntactical checks.

`verifyta`: A model-checker that combines *on-the-fly* verification with constraint solving techniques [**YPD94, LPY95a**].

During the past year, we have applied UPPAAL to a number of case-studies. To meet requirements arising from the case studies, the UPPAAL model and model–checker have been further extended with a number of new features:

Committed Locations. UPPAAL adopts hand-shaking synchronization between components in a network. A very recent case-study on the verification of Philips Audio Control Protocol with bus-collisions [**BGK$^+$96**] shows that we need to further extend the UPPAAL model with *committed locations* to model behaviors such as atomic broadcasting in real-time systems. The notion of committed locations is introduced in [**BGK$^+$96**]. Our experiences with UPPAAL show that the notion of committed locations implemented in UPPAAL is not only useful in modeling real-time systems but also yields significant reductions in time- and space-usages in verifying such systems.

Urgent Actions. In order to model progress properties UPPAAL uses a notion of *maximal delay* that requires discrete transitions to be taken within a certain time bound. However, in some examples, e.g. the Manufacturing Plant [**DY95**], synchronization on certain channels should happen immediately. For this reason the UPPAAL model was extended with *urgent channels*, on which processes should synchronize whenever possible [**BLL$^+$95**]. The notion of urgent channels (also known as *urgent actions* in the literature) has been implemented in both HYTECH and KRONOS.

Diagnostic Traces. Ideally, a model-checker should be able to report diagnostic information whenever the verification of a particular real-time system fails. UPPAAL reports such information by generating a *diagnostic trace* from the initial state to a state violating the property. The usefulness of this kind of information was shown during the debugging of an early version of Philips Audio-Control Protocol [**LPY95b**].

UPPAAL has been applied to a number of case-studies and benchmark examples, including: several versions of Fischers Protocol [**AL93**], two version of Philips Audio-Control Protocol [**BPV94, LPY95b, BGK**$^+$**96**], a Steam Generator [**Abr95**], a Train Gate Controller [**HHWT95**], a Manufacturing Plant [**DY95**], a Mine-Pump Controller [**JBW**$^+$**96**] and a Water Tank [**OSY94**].

The growing list of succesfully completed real–size verification case–studies and recently initiated collaboration with danish industry makes us believe that the UPPAAL is reaching a level of maturity where it can be applied to real industrial case–studies.

3.2. The UPPAAL **design.** Having already made the PROMELA protocol design actually made the modelling using timed automata a relatively easy task. The time spent on the timed automata design has been considerably less than the time spent on the initial PROMELA design.

The fundamental automata design is quite similar to the PROMELA design in the sense that the same type of processes are modelled. That is, each PROMELA process is matched by a timed automaton in the new design. The main differences in the models, besides the timing, considers the way communication between processes takes part and the way the broadcasting behaviour of the medium is expressed. In the timed automata model used in UPPAAL there is no channel primitive and the only means of interaction between automata is by pure synchronization of atomic actions. Consequently, we use a combination of shared variables and synchronization to simulate the message passing that would actually take place in a real system.

The master. The master process is modelled as the timed automaton depicted in Figure 8[4]. We consider a lossy communication medium and therefore the master is equipped with a timer, see Figure 9, to guarantee that new enquiries will be sent in the precense of message loss.

The master starts by sending to the medium an enquiry addressed to the first slave. This is modelled by the initial transition from `m0` to `m1`. The master sets the shared variable `data:=0` on this transition indicating that the message is an enquiry. Having sent the enquiry and without further delay, the master sets its timer and starts waiting until the message has been broadcasted to all slaves, indicated by an `empty` synchronization with the medium. This ensures that the master will not receive its own message.

Now, the master will either receive data broadcasted by a slave or it will time-out, if nothing is received within a certain time limit. In either case the master sends an enquiry along to the next slave indicated by increasing the shared variable `next`, setting `data:=0` and performing the output action `to_medium!`.

The Medium. As mentioned the basic means of interaction between timed automata is by binary synchronization. No basic broadcasting primitive exists, but the notion of committed locations, see section 3.1, can be used to model the broadcasting in a simple way. Having received a message by synchronizing on the input action `to_medium?`, the medium delays the message for one time unit and then it starts broadcasting, see Figure 10. The broadcast consists of synchronizing in turn with each of the not–sending processes connected to the medium. Atomicity of the synchronization sequence is ensured by labelling each node that participates

[4]This figure shows the actual input to UPPAAL.

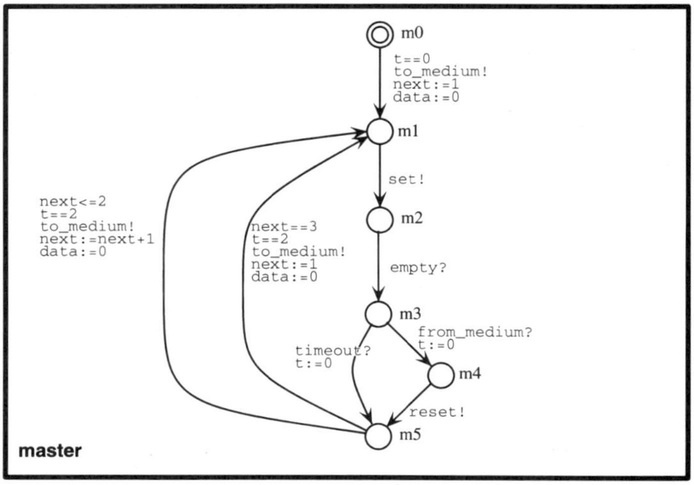

Figure 8: The Master with Timer

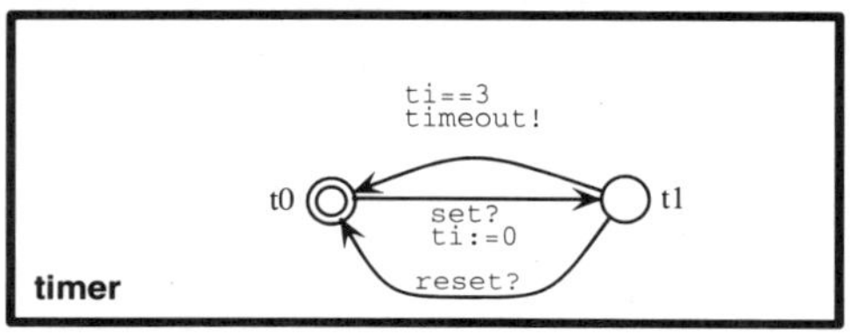

Figure 9: The Timer

in the synchronization sequence as committed. This guarantees that no actions can interleave the broadcast.

The node labelled `col` will be entered upon collisions in the medium, and it serves the same verification purpose as the `accept_collision` state in the PROMELA model, see Figure 4.

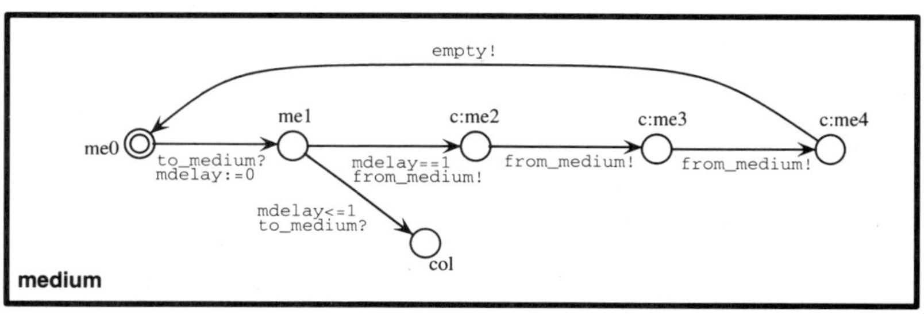

Figure 10: The Medium

The Slaves. In the UPPAAL model we need to model each slave as a unique timed automaton. In Figure 11 one of the almost identical slaves is depicted. The slaves synchronize with the medium on input action `from_medium?` and either they loose messages or they receive correctly, in which case they now determine what type of data is sent and to whom. Depending on the outcome, slaves either return to their initial state, sends data along to their users or asks their users for data to be send. In the last two situations the slaves will delay some amount of time and during this period they will not be able to detect messages sent to them. This is modelled as the 'ignoring' **from_medium?** input actions at the nodes **s1_2** and **s1_4** of Figure 11.

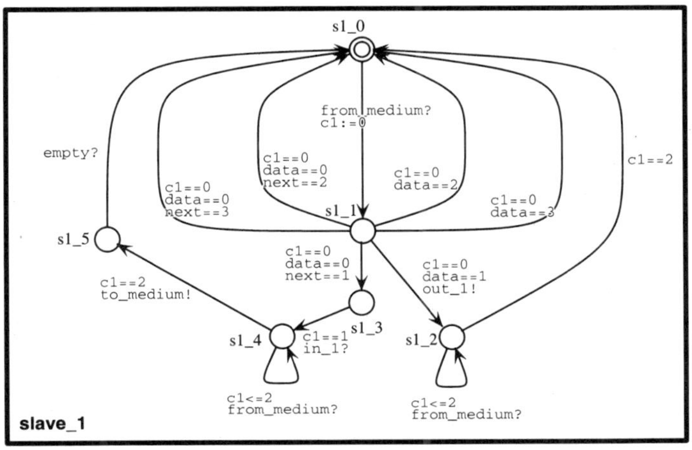

Figure 11: A Slave

The Users. As for the slaves we need to model each user process as a unique timed automaton. In Figure 12 the user automaton of the slave in Figure 11 is depicted. Users are always ready to either responding to enquiries from their slaves or receiving data sent from other users. Responding to enquiries is done by sending data to another user. The committed locations in the user automaton are for verification purposes and will be explained in section 3.3.

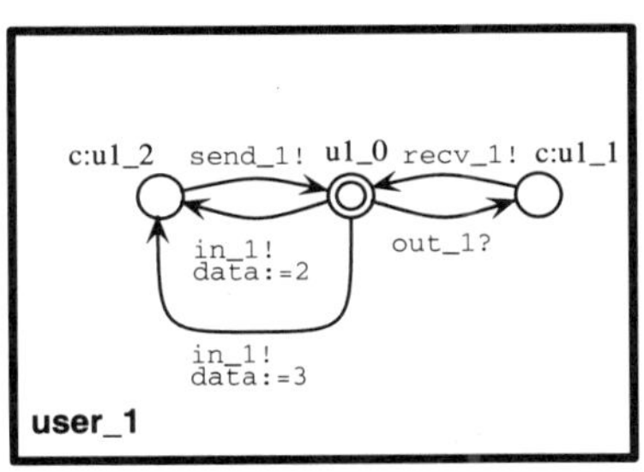

Figure 12: A User

3.3. Verification in UPPAAL. The primary correctness criteria that we want to verify for the protocol design explained in section 3.2 is that no collisions will ever occur. As the medium delays messages for one time unit, two messages sent to the medium within one time unit or less will eventually collide and this scenario will force the medium automaton in Figure 10 in the node `col`. What we need to verify is that it holds invariantly, that the protocol can not reach a state where the medium automaton is in state `col`. Stated as a property in the logic of UPPAAL this becomes:

$$\forall\Box(\texttt{not medium.col})$$

The satisfaction of the above formula is dependent upon the actual timeout limit in the timer. UPPAAL succesfully verifies the property if we consider a perfect medium, i.e. not lossy. But when an erroneous medium is introduced as in section 3.2 the timeout limit influences the possibility of collisions. If a timeout occurs to soon, the master interprets this as a situation where data is lost and all slaves are waiting for messages. But obviously this need not be the case as a slave can actually be in the process of enquiring its user. If this happens the slave will try to send data from its user and the master will try to send a new enquiry. If these two messages arrive at the medium within the one time unit delay of the medium, they will collide. We discover by repeated verification attempts that timeout limits greater than or equal to 3 will ensure that no collisions can occur. Also we verify that for a timeout limit of 2, a collision actually can occur, and the diagnostic trace facility of UPPAAL gives us a possible trace leading to collision.

Assuming a perfect medium (not lossy) and assuming that data is sent from users in round–robin fashion (all user are interested) we want to verify that the user–to–user communication delay is bounded by some constant. Also, we want to verify an upper bound on the delay between users sending data. This actually implies a bound on the delay between enquiries from the master, as all users are interested in sending. The above properties are examples of *bounded liveness* properties which can not be expressed directly in the logical property language of UPPAAL. To express the properties we introduce a seperate test automaton that probes the user processes in the protocol design. The test automaton will be designed to enter a 'bad node' if it tests an unwanted behaviour of the protocol. This approach is quite analogous to the never–claims used in the PROMELA language.

The test automaton for the properties described above is depicted in Figure 13. The automaton probes the sending and receiving of data in the user processes by synchronizing on actions `send_1` and `recv_1` (for user 1), see Figure 12. When a message is sent the test clock `s` is started and if the data sent is not received within a certain time limit, the test automaton is forced in the state `bad1`. Similarly, if a new sending is not performed within a certain time from the last receiving, the bad state `bad2` can be entered. The property verified using UPPAAL is:

$$\forall\Box\ \texttt{not (check\_1.bad1 or check\_1.bad2)}$$

Using a similar approach as above we verify that there exists a round–trip time bound for the protocol. We use the test automaton of Figure 14 to verify that there exists a round–trip, modelled as user 1 having performed two sends, within a certain time bound. We verify:

$$\exists\Diamond(\texttt{check\_2.ch2 and s} \leq 18)$$

Also we verify that the following does not hold:

$$\exists\Diamond(\texttt{check\_2.ch2 and s} \leq 17)$$

That is, there exists no (initial) round–trip time of less than 18 time units.

As indicated on Figure 8 the master waits for two time units before sending out a new enquiry. This time limit guarantees that all slaves have finished their internal

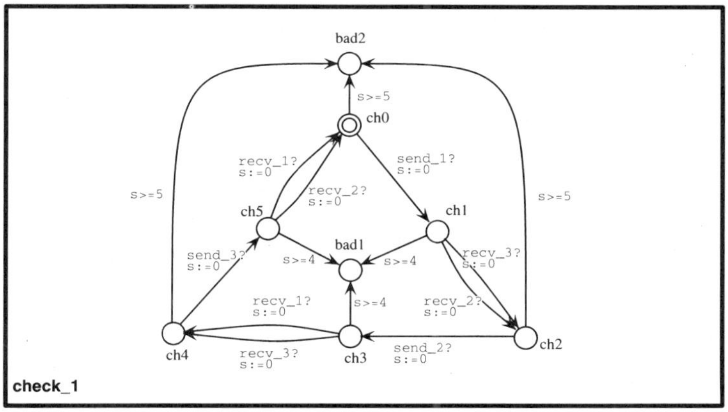

Figure 13: Bounded liveness test automaton

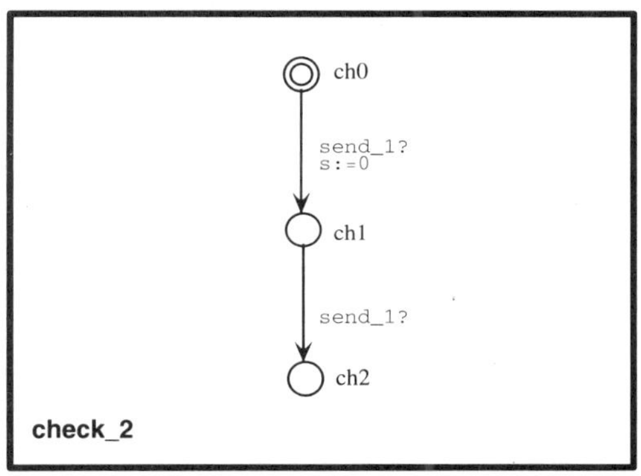

Figure 14: The round–trip time test automaton

buissness and will be ready to receive data. I.e. messages will not be lost because of slave not ready to receive. To increase the round–trip performance of the protocol we could consider to let the master send out new enquiries without delaying at all. Obviously this would require redesigning the enquiring strategy of the master s.t. it will not send enquiries to slaves having just received data from other users. We believe that a new strategy for the master benefiting from the parallelism in the slaves, will be possible, but we have not considered the actual design. However, we have verified that changing the waiting time in the master to zero, will allow for faster round–trip times. Obviously changing only the waiting time of the master does not guarantee that messages will not be lost, but it gives a clue as to how the enquiring strategy can be optimized.

3.4. Test Automata Generation. In the previous section we verified a number of bounded liveness properties by establishing reachability properties in the context of a testing automata. To allow the testing automata to 'observe' the system via communication, the system was in most cases extended with suitable probe actions.

Similar to SPIN's ability to generate never–claims directly from Linear Temporal Logic properties, it is possible to derive testing automata automatically from logical properties of the Safety and Bounded Liveness Logic introduced and studied in [**LPY95a, LPY95b**]. Here we indicate the automatic derivation of testing automata for a somewhat simpler logic STL (Simple Timed Logic) derived from the logic TML introduced in [**HLY92**].

The properties of STL is given by the following abstract syntax, where a ranges over actions and N over natural numbers (extended with ∞) :

$$\varphi \ ::= \ \mathrm{tt} \mid \mathrm{ff} \mid \mathrm{INV}\varphi \mid \langle a\rangle_{\leq n} \mid [a]_{\leq n}\varphi \mid \varphi_1 \wedge \varphi_2$$

The properties of STL are interpreted with respect to the behaviours (i.e. timed transition systems) of (networks of) timed automata.

The interpretation of the propositional part is standard, and $\mathrm{INV}\varphi$ requires as expected that any reachable state of the timed automata must satisfy the property φ. The time–quantified action modality $\langle a\rangle_{\leq n}$ describes informally that the system must be able to perform an a–action no later than (the observer has experienced) a delay of n [5]. Similarly, $[a]_{\leq n}\varphi$ requires that any a–transition occurring before a delay of n must lead to a new state satisfying φ. We write $A_1 \mid \ldots \mid A_n \models \varphi$, when a network of timed automata, $A_1 \mid \ldots \mid A_n$ satisfies an STL formula φ.

Now, for any STL formula we may construct a testing timed automata T_φ with a designated location l_φ such that

$$l_\varphi \text{ is } \textit{un}\text{reachable in } \big(T_\varphi \mid A_1 \mid \ldots \mid A_n\big)$$
$$\textit{if and only if}$$
$$\big(A_1 \mid \ldots \mid A_n\big) \models \varphi$$

Thus model–checking STL properties may be reduced to deciding reach-ability questions. The testing automata T_φ is defined by the structure of φ and given in Figure 15.

In section 3.3 we essentially wanted to check that a sender can not send messages too frequently: at least 18 time units between two consecutive send's must elapse. The following property:

$$\mathrm{INV}\big([send_1]_{\leq\infty}[send_1]_{\leq n}\mathrm{ff}\big) \tag{3.1}$$

expresses precisely that at least n time units must elapse between two consecutive $send_1$ actions. Using the constructions described above we obtain the testing timed automata of Figure 16. That is we may check for the property (3.1) by checking for reachability of location l when the testing automata is combined with the system under consideration. The testing automata that was used in the actual verification was a slight simplification of the one that is obtained by the general construction.

4. Comparison of SPIN and UPPAAL

In this section we summarize our experiences with the two tools SPIN and UPPAAL based on our experience from the case study concerning the Collision Avoidance Protocol for an Ethernet.

[5] The observer experiencing a delay of n means that the system may perform a sequence of delay steps intermixed with internal computations such that the delay steps accumulate to a total of n.

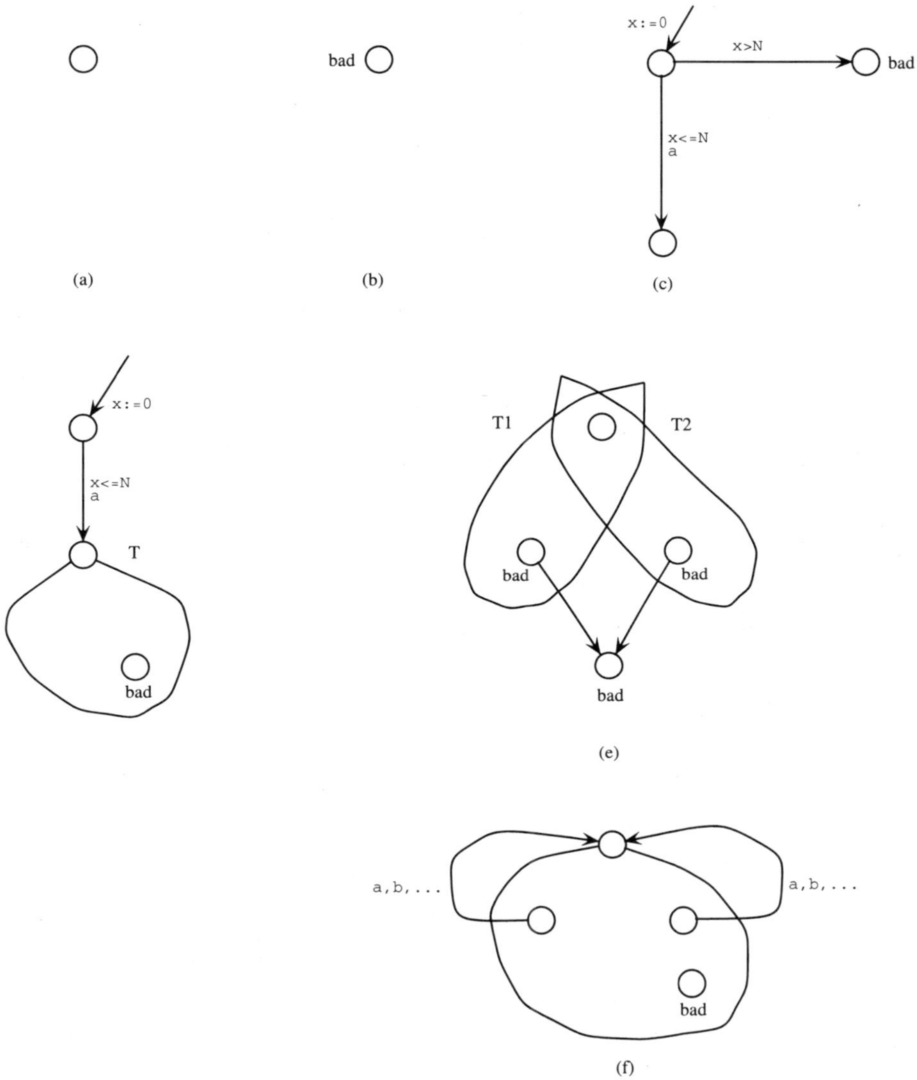

Figure 15: Generation of Testing Timed Automata: The respective test automata implements the following formulas: `(a)` tt, `(b)` ff, `(c)` $\langle a \rangle_{\leq n}$, `(d)` $[a]_{\leq n}\varphi$, `(e)` $\varphi_1 \wedge \varphi_2$ and `(f)` INVφ. In the figure, `T` indicates the testing automaton for φ. `T1` and `T2` indicates the automata for φ_1 and φ_2 respectively.

Considering the design phase, the basic structure of the design was very easily obtained in the UPPAAL model and the PROMELA/SPIN model was of great benefit as a structural basis for our UPPAAL model, in the sense that each PROMELA process was matched by a timed automaton in the UPPAAL design.

In the PROMELA design phase we made extensive use of the simulation facilities of SPIN, especially the Message Sequence Charts. Within short time a 'running' prototype was designed and at an early stage faults were detected without having

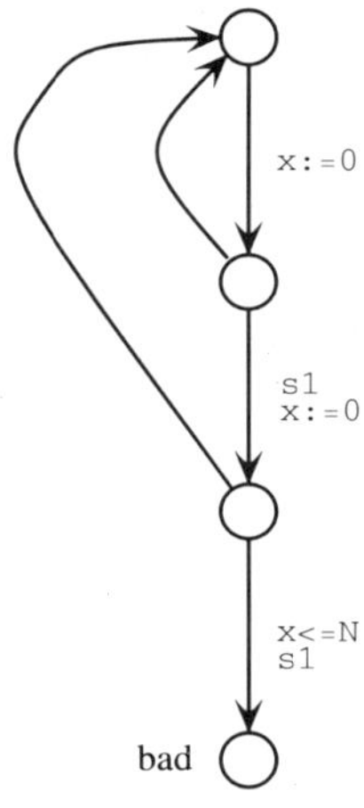

Figure 16: Test Automata for the property of (3.1)

the full design at hand. In contrast UPPAAL does not yet allow for simulations and consequently, the UPPAAL design has to be more fully developed 'by hand' before automatic tool support can be applied. This means, that the benefits of tool support in the UPPAAL design process are delayed compared to the SPIN design process.

Considering the design languages, the obvious distinction is the possibility of modelling real–time systems in UPPAAL. In the case study it is shown that interesting bounded liveness properties can be expressed and verified in UPPAAL. Another beneficial feature of UPPAAL is the possibility of committed locations which makes possible a quite natural modelling of the broadcast behaviour needed in the case study. In contrast PROMELA can not apply the atomicity construct on sequences of send- and receive statements as these might be blocking.

Considering the verification phase, the kind of properties expressible in the property language of UPPAAL are restricted to invariance and possibility properties. Other properties as e.g. the bounded liveness properties of our case study needs to be expressed as separate test automata probing the design. In section 3.4 we present ideas on how to extend the property language and automatically generate the test automata. This is already possible in SPIN for transforming LTL properties to never automata.

The committed locations of UPPAAL make it possible to design non realizable systems. In particular systems that may enter completely blocked states (in the sense that neither actions nor time delays are possible) can be described. Obviously, we would like the possibility of checking whether the global design suffers such unrealizable properties or not.

Both SPIN and UPPAAL offers diagnostic information upon negative verification results. SPIN offers the possibility of examine an error scenario using the MSC's and UPPAAL offers a textual sample error trace leading to the unwanted state. By performing breadth first reachability analysis UPPAAL makes available a shortest error trace, whereas this is not guaranteed in SPIN as the reachability is performed depth first.

References

[Abr95] J.-R. Abrial, *Steam-boiler control specification problem*, International Seminar on Methods for Semantics and Specification (1995).

[AL93] Martin Abadi and Leslie Lamport, *An Old-Fashioned Recipe for Real Time*, Lecture Notes in Computer Science **600** (1993).

[BGK+96] Johan Bengtsson, David Griffioen, Kåre Kristoffersen, Kim G. Larsen, Fredrik Larsson, Paul Pettersson, and Wang Yi, *Verification of an Audio Protocol with Bus Collision Using* UPPAAL, Accepted for presentation at the *8th* Int. Conf. on Computer Aided Verification, 1996.

[BLL+95] Johan Bengtsson, Kim G. Larsen, Fredrik Larsson, Paul Pettersson, and Wang Yi, UPPAAL— *a Tool Suite for Automatic Verification of Real–Time Systems*, Proc. of the *4th* DIMACS Workshop on Verification and Control of Hybrid Systems, Lecture Notes in Computer Science, October 1995.

[BPV94] D. Bosscher, I. Polak, and F. Vaandrager, *Verification of an Audio-Control Protocol*, Proc. of FTRTFT'94, Lecture Notes in Computer Science, vol. 863, 1994.

[DY95] C. Daws and S. Yovine, *Two examples of verification of multirate timed automata with* KRONOS, Proc. of the *16th* IEEE Real-Time Systems Symposium, December 1995, pp. 66–75.

[HHWT95] Thomas A. Henzinger, Pei-Hsin Ho, and Howard Wong-Toi, *A Users Guide to* HYTECH, Tech. report, Department of Computer Science, Cornell University, 1995.

[HLY92] U. Holmer, K.G. Larsen, and W. Yi, *Decidability of bisimulation equivalence between regular timed processes*, Proc. of CAV'91 (Berlin), Lecture Notes in Computer Science, Springer Verlag, vol. 575, 1992.

[Hol] Gerard J. Holzmann, *Basic spin manual*, AT&T Bell Laboratories, Murray Hill, New Jersey.

[Hol91] Gerard J. Holzmann, *Design and validation of computer protocols*, Prentice Hall, 1991.

[JBW+96] Mathai Joseph, Alan Burns, Andy Welling, Krithi Ramamritham, Jozef Hooman, Steve Schneider, Zhiming Liu, and Henk Schepers, *Real-time Systems Specification, Verification and Analysis*, Prentice Hall, 1996.

[KV96] K. Karsisto and A. Valmari, *Verification-driven development of a collision avoidance protocol for the ethernet*, FTRTFT96 (1996).

[LPY95a] Kim G. Larsen, Paul Pettersson, and Wang Yi, *Compositional and Symbolic Model-Checking of Real-Time Systems*, Proc. of the *16th* IEEE Real-Time Systems Symposium, December 1995, pp. 76–87.

[LPY95b] Kim G. Larsen, Paul Pettersson, and Wang Yi, *Diagnostic Model-Checking for Real-Time Systems*, Proc. of the *4th* DIMACS Workshop on Verification and Control of Hybrid Systems, Lecture Notes in Computer Science, October 1995.

[OSY94] A. Olivero, J. Sifakis, and S. Yovine, *Using Abstractions for the Verification of Linear Hybrids Systems*, Proc. of CAV'94, Lecture Notes in Computer Science, vol. 818, 1994.

[TC96] Stavros Tripakis and Costas Courcoubetis, *Extending promela and spin for real-time*, Tools and Algorithms for the Construction and Analysis of Systems, Second International Workshop, TACAS '96, Lecture Notes in Computer Science, vol. 1055, 1996, pp. 329–348.

[YPD94] Wang Yi, Paul Pettersson, and Mats Daniels, *Automatic Verification of Real-Time Communicating Systems By Constraint-Solving*, Proc. of the *7th* International Conference on Formal Description Techniques, 1994.

DEPARTMENT OF MATHEMATICS AND COMPUTER SCIENCE, AALBORG UNIVERSITY, DENMARK
E-mail address: `ejersbo@cs.auc.dk`

DEPARTMENT OF MATHEMATICS AND COMPUTER SCIENCE, AALBORG UNIVERSITY, DENMARK
E-mail address: `kgl@cs.auc.dk`

DEPARTMENT OF MATHEMATICS AND COMPUTER SCIENCE, AALBORG UNIVERSITY, DENMARK
E-mail address: `ask@cs.auc.dk`

DIMACS Series in Discrete Mathematics
and Theoretical Computer Science
Volume **32**, 1997

The Application of PROMELA and SPIN in the BOS Project

Pim Kars

ABSTRACT. The BOS project concerns the design and implementation of a storm surge barrier control system. The SPIN tool set (SPIN and XSPIN) was used to validate parts of the design, in particular the communication interfaces with the outside world. Furthermore, PROMELA combined with Z is used to specify crucial aspects of the design. In this paper we relate our experience with PROMELA/SPIN in the project.

After a short introduction to the BOS project in section 1, we discuss the "why, what and how" of the use of formal methods in the project in section 2. Our most recent experience in validation is treated in section 3. Most papers on validation tend to discuss either theory (algorithms, logics) or case studies showing how a system was modelled and validated. This leaves some interesting issues underdeveloped: the pragmatics of validation using PROMELA/SPIN. We want to share some preliminary reflections on these issues in section 4. Finally we present some ideas for extensions in section 5, and concluding remarks in section 6.

1. The BOS project

After the last major storm flood of the low, western part of the Netherlands in 1953, a project was launched by the Dutch Ministery of Public Works to prevent such floods from happening in the future. The keystone of this 40-year project is the construction of a movable storm surge barrier near the Hook of Holland in the Nieuwe Waterweg, the canal connecting the harbour of Rotterdam to the North Sea.

The barrier will operate fully automatically because human intervention was deemed too error prone. BOS, an acronym for "Beslis- en Ondersteunend Systeem" (Decision and Support System), is the central system that decides when and how the barrier has to move, and that controls its movements. In order to perform these tasks, the BOS system collects measurements of water levels, tidal info, weather data and predictions from several sensors and networks. Furthermore, it includes a software module (provided by a third party) that predicts water levels based on the various measurements.

The BOS system has to satisfy very high reliability requirements expressed as failure rates of undesirable events. For example, the probability of failure to close

1991 *Mathematics Subject Classification.* Primary 68Q60; Secondary 68N99, 68U30.

the barrier when called for should be less than $3 * 10^{-4}$, and the probability of failure to open the barrier in time after closure should be less than $5 * 10^{-5}$. Not closing in time could result in the flooding of cities like Rotterdam. If the barrier is not opened in time after a closure, it could be severely damaged. Moreover, while the barrier is closed, the important harbour of Rotterdam is cut off from the North Sea, leading to great economic loss.

Computer Management Group (CMG) — a company providing a wide range of information technology services with branches in the United Kingdom, The Netherlands and Germany — was awarded the contract from the Ministery of Public Works to build BOS. It was clear from the outset to project management at CMG that extra measures were needed in this project given the high reliability requirements, because it is infeasible to quantify reliability with such precision [**BF91**]. The IEC 65A draft standard for highly reliable systems (and its successor, the IEC 1508 standard) strongly recommends the use of formal methods. Therefore, CMG set up a close collaboration with the Formal Methods group at the University of Twente to transfer knowledge and provide consultancy on the application of formal methods to system design. (As an aside, the use of "we" in this paper therefore generally includes both the BOS project team at CMG and the Formal Methods group at the University of Twente.)

2. Formal methods in the BOS project

2.1. Choice of methods. One should think carefully about why, what and how to use formal methods in a project. It is not feasible in practice to formally specify a large system and, for example, use refinement techniques to get to an implementation. Another reason to be cautious about the goals one has with introducing formal methods into an existing design trajectory, is that it is risky to introduce too many new techniques at one time into a project.

As part of an effort to integrate formal methods within CMG's design trajectory, the decision was made to concentrate on two tasks:

- validate crucial parts of the design, in particular the interfaces with the outside world, to get more confidence in the correctness of the mechanisms chosen in the design, and
- formally specify the design, primarily to avoid ambiguities and incompleteness.

As to the choice of methods, for validation PROMELA/SPIN was a natural choice: we already had experience using SPIN, it is able to handle large state spaces, and the language and tool are relatively easy to learn and use. The last factor was particularly important since the desire to integrate formal methods meant that engineers at CMG would do most of the actual work. The choice of a specification language for the design was harder. One of the limiting conditions was the ability to specify control as well as data aspects. No language that we had experience with completely fitted the bill. The solution was to use two languages: PROMELA (which already had to be learnt for the validation) for the control part and Z [**Spi92**], embedded in PROMELA, for the data part.

2.2. Validation. SPIN was used to validate most of the communication interfaces with the outside world and parts of the internal process control layer. A typical validation model has three (or more) processes: one process to model the interface, one process to model the environment, and one or more "user processes". The environment process typically consists of a non-deterministic choice between normal actions and error actions/failure modes (losing a message, spontaneous responses due to noise, etc.). The user processes encode the requirements on the expected service provided by the combination of the interface and the environment.

We used an incremental approach for validation, which seems fairly standard.

1. Start with a design and a friendly environment, i.e. without error actions.
2. Iterate validation cycle
    - IF violations found: fix the design.
    - ELSE add more error behaviour to the environment.

A validation cycle starts with checking simple properties, then on to the more expensive properties. Each new design is first simulated to check that it correctly models the intentions of the designer.

2.3. Specification. Besides the communication interfaces, also other parts of the design are specified using a combination of PROMELA and Z. Z schemas are used to model data (in particular the state of a subsystem), and also to specify operations on the state. The Z operations are embedded in a PROMELA program to add control. We adopt the convention that a Z operation is executable when its precondition holds. This embedding is ad hoc in that no formal semantics is given for the combination. This can only be done when PROMELA is provided with a formal semantics.

The advantage of this approach is that a powerful, expressive language is used to specify data and operations. A potential drawback is that Z is not constructive and no tools exist to validate the combination. However, the specification is primarily intended to give an unambiguous specification of what has to be built. Also the Z *style* that is used is largely constructive, so that e.g. test oracles might be derived from the Z specifications. Z type checkers, notably `ztc` and `fuzz`, are used to check the consistency of the Z parts of the specification.

2.4. Integration. Formal methods are only one means to enhance the quality of the system. As such they should be integrated with other measures. Starting from a functional specification, the strategy to get to an implementation was set up roughly as follows:

1. Basic design: decomposition of the functionality into subsystems,
2. Function Failure Analysis to determine critical functions,
3. Detailed design per subsystem:
    - specification in Z and PROMELA
    - validation of critical parts with SPIN
    - selective prototyping
    - review of the design
4. Implementation in C++ and extensive testing by separate teams.

At the moment, the design is almost finished and implementation is under way. The detailed design is described in a set of documents, one per subsystem, containing

- Dutch text for descriptions and explanations,
- PROMELA for specifying I/O and control structure,
- Z for specifying data and operations on data, and
- Data Flow Diagrams for documenting the overall structure.

A case tool SDW [**Cap93**] is used to link the elements in order to maintain consistency and the relation to the Functional Specification.

2.5. Experience. On the whole our experience with SPIN and PROMELA was positive.

- SPIN/PROMELA is very easy to learn and use, even for an engineer with no previous exposure to formal methods. This was important since most of the actual work was done by engineers of CMG. Most of the language elements are simple and intuitive, there is a fine tutorial, and many of the validation requirements can be expressed in a simple way without having to learn temporal logic.
- SPIN was useful to expose errors in the proposed communication interfaces and in getting confidence in the final design.
- Message Sequence Charts were a valuable aid in convincing management of problems in the protocols and the need to repair them.

The use of Z turned out to be less simple than expected. More on this subject will be reported elsewhere; see also [**Wij96**].

3. Recent experience

3.1. BBI, the BOS-BESW interface. Our latest endeavour was the validation of one of the most crucial interfaces: the interface between BOS, the system that takes the decision to move the barrier, and BESW, the system that performs this task. BESW is an acronym for "BEsturingsSysteem Waterweg" (Control System Waterweg). It consists of a separate network of controllers (provided by a third party) that control the actual movements of all parts of the barrier.

The interface between BOS and BESW is called BBI. Originally this interface was designed as a simple Question-Answer service built on top of a reliable, duplicated data link service, the details of which were left to be filled in later. BOS, acting as the master, sends a command to BESW and BESW answers with an acknowledgement, or BOS sends a status request and BESW answers with the requested status. Apart from reliably transmitting commands, status requests and replies, the interface also provides a "heartbeat" mechanism such that failure of the BOS system or both links to it can be detected by BESW. If such an unlikely event occurs, BESW switches to manual operation. When the time came to fill in the details a setup was chosen in which BOS remotely accesses registers within BESW: bits in these registers are set to signal events.

3.2. Validation of the BBI. We will not go into much technical detail of the interface and the model, but concentrate on observations of the validation process.

It was clear from the start that the interface would not be free of errors. In fact, some recent changes in the requirements had made the design incomplete at some points. So our task was not to try to show that it was correct, but rather to find enough severe errors to invalidate the design.

To be able to model the system for validation, specifications of both sides of the interface were needed. The BOS side of the interface was well-documented in a mixture of PROMELA and Z augmented with informal explanations. The parts relevant to the validation could be translated into PROMELA with rather small effort. The documentation of the BESW side of the interface was harder to interpret; some but not all of the scenarios were explained by a kind of timing diagrams. Analysis of the BESW side uncovered several unspecified situations, mostly to do with combinations of events (potential race conditions), that had to be resolved. Subsequent discussions with the company building BESW showed that indeed these situations would lead to unwanted behaviour including deadlocks. Also enough details of BESW's behaviour became known to finish the model and start simulation and validation.

It is interesting to note that most errors were not found during validation runs but during this preliminary analysis stage. Some errors were found by simulation and only a few errors during validation. The whole enterprise has been successful in the sense that sufficient errors were found to convince management that the interfaçe should be redesigned, a process that is currently taking place.

3.3. Some modelling issues in the BBI model. The processes relevant to the BBI were straightforwardly modelled as proctypes connected through asynchronous channels. The processes typically operate in cycles where each cycle consists of a choice between `atomic` steps. The BOS side was kept as close to the specification as possible to minimize translation errors. However some "expensive" mechanisms were simplified. For instance, the BOS process directly connected to the BOS side of BBI uses polling requests to wait for completion of an issued command. In the model this was replaced by waiting for a variable to reach a certain value.

Since PROMELA has no direct support for modelling time (apart from explicitly modelling clocks) the BBI model was made asynchronous. An advantage of this approach is that the model is less sensitive to deviations in timing. A major drawback however is that the model allows behaviour that "in reality" cannot occur. So each time an error is reported, one has to go back to the original design and check whether the error is really an error in the design or merely a consequence of abstracting from time. Many of the errors reported turned out to belong to the second category, especially when simulation was used. Several solutions can be envisaged to this problem:

- use a timed extension of PROMELA like [**TC96**],
- constrain the behaviour by introducing extra synchronizations between processes, e.g. by handing out tokens to regulate the relative speed of processes, using provided clauses to proctypes or using timeout to implement clock cycles, or
- constrain the behaviour in the never claim (only applies to validation, not to simulation).

In the BBI model we needed to model a register with several bits, say 8 of them. A simple way to express this in PROMELA is to use an array of bits:

```
bit command_reg[8] ;
command_reg[5] = 1
```

However, a more compact representation is obtained using plain bytes to model registers and bit-operations on bytes:

```
byte command_reg ;
command_reg = command_reg | (1 << 5)
```

Experiments showed that the second representation not only uses less memory, it also runs faster.

In many cases looping through the bits of a register was best considered as a new primitive action modelled as a `d_step`. Unfortunately this was not always possible since SPIN currently does not allow a `d_step` within an `atomic` sequence.

Timeouts on asynchronous channels are usually modelled "locally" by checking for an empty channel:

```
if
::  channel?data -> ...
::  empty(channel) -> ...  timeout behaviour ...
fi
```

Modelling timeouts in this way gives smaller models than using `skip`. Premature timeouts are still covered in our models due to the interleaving of sender and receiver processes.

4. Reflections on SPIN pragmatics

Some case studies give the impression that validation is just a matter of translating a system into a model and push a button; if the result is OK then go ahead and implement the system, otherwise fix the design error and reiterate. An underlying assumption seems to be that it is clear what the system and validation criteria are; that they are given and fixed once and for all. Obviously this is something to strive for, but it is our experience that in reality things are not as simple as that. Gerard Holzmann [**Hol91**] gives some advice on how to model for SPIN validation. We would like to add some comments based on our experience, and outline some of the problems a validator faces when using PROMELA/SPIN (or probably any validation system for that matter).

4.1. The validation process. To validate a design, be it with SPIN or another tool, several steps must be taken:

1. formalize the design, i.e. translate it into the modelling language,
2. verify the formalization with respect to the intention,
3. formalize the validation criteria, i.e. translate them into the validation language (assertions in monitor processes, PROMELA never claims, LTL formulae),
4. validate the design against the criteria.

It is often assumed that step 2 is best done by simulation and the last by validation. However, there are cases where the two cannot really be separated. In fact, each time a validation run reports an error, the cause may either be in the model (modelling error) or in the original design. Just as the software development process is better described by the spiral model than the waterfall model, here also the steps are not to be considered as discrete steps followed sequentially, but more as a cyclic activity.

4.2. Formalization of the design. It may be useful to first indicate the kind of systems we have experience with. The systems we have been validating are mostly interfaces of the BOS system with the outside world. Typically three (or more) entities can be discerned:

- the subsystem being modelled,
- the environment (part of the outside world, i.e. outside of the BOS system), and

- the inner environment: the rest of the BOS system as far as relevant for the validation.

These entities can usually be faithfully modelled as processes communicating asynchronously. The subsystem being modelled is already formally specified in Z and PROMELA and ideally the model is kept close to this specification. The environment has to be formalized based on documents provided by the supplier. The inner environment has to be abstracted from the relevant specifications. Of these three, the subsystem is really the model in a strict sense. The other two entities are only there to "drive" the model of the subsystem; it is often useful to combine them into a common "driver" process to guide and check end-to-end behaviour, cf. the notion of an upper tester.

There are several circumstances that complicate the formalization such as

- the environment may have several failure modes,
- the behaviour of the environment may not be fully specified, requiring further analysis to complete its specification, and
- the inner environment may be scenario-based.

Modelling issues: structure and use of language elements. Usually there is more than one way to express something in a language, be it a specification language or a system implementation language, and PROMELA is no exception to this rule. (If some aspects of the system cannot be represented in the modelling language, an abstraction has to be made rightaway.) As an example see the discussion on modelling registers in section 3.3. As in system implementation languages, issues here seem to be clarity and flexibility versus compactness of code and speed of execution.

It is a good idea to start thinking about how to include the failure modes in the beginning. This means that the model should be structured in such a way that changes in the failure behaviour are easily made.

Also, it is important that the model is structured in such a way that events important to the validation are readily observable. For example these could be read or write events in channels, the change of a global variable (should be global, otherwise it is not observable to a never claim) or reaching a certain state label. These events and their representation should be decided upon early on, because the expression of validation criteria depends on them.

Level of abstraction. Preferably, the model is kept as close to the original system specification as possible to minimize translation errors. As is well known, however, a complex system may be expensive to validate. Therefore a key issue in modelling a system for validation is to reduce the complexity by abstracting away from irrelevant details (see e.g. [**Hol91**, section 11.8]). But what are irrelevant details? This depends on the part of the system one is interested in and the (kind of) errors one is looking for. By the way, any validation effort in the SPIN (and others) fashion can be described as a search for errors of an expected kind; errors of an unexpected kind will not be revealed.

Also the link with the original system needs to be maintained. At one extreme, abstracting away too much detail may result in a model with no errors and hardly any relationship with the original system to be validated. More remarks on this issue follow below.

Summarizing, we have the following constraints on the model:

1. it should have a close, known relation to the intended system; preferably this relation should also be explicit, well-documented,
2. it should be concrete enough to model all system peculiarities, and
3. it should be small enough to validate in an acceptable time.

Traceability. A serious problem is how to maintain the link between the (possibly informal) system to be validated and the model that is eventually validated. When errors are found in the model, the original design may be adapted, or the model is adapted to accomodate to the whims of the modelling language, etc. A first observation is that the development of a model may thus be compared to the development of code. In a serious software development project, where the *traceability* of developments and decisions is of utmost importance, one should seriously consider using a version control system. In any case, it is useful to keep a log-book of the validation activities just as one would do in laboratory experiments.

Secondly, this seems to demand an approach where the system is specified in a rather concrete fashion, and next this concrete specification is used both for validation and implementation:

- reduce/project the specification for validation,
- refine the specification for implementation.

A reduction strategy, which in our experience can be useful, is to project on (i.e. "hide") syntactical entities like variables, processes, mtypes and channels. A simple strategy could just replace these entities by "don't care" entities; a more refined one would have to use some kind of abstract interpretation, an area of much current research. Tool support for these kind of reductions under user guidance would be most welcome. This is an alternative to manual reduction and proving that the result is equivalent to the original, for example using a tool like `spine` [**Erd95**]. Or to reduction strategies implemented *inside* the tool such as partial-order reduction as already implemented in SPIN (based on the class of properties expressible in nexttime-free linear temporal logic) or dedicated reduction strategies based on the specific property that is being checked.

4.3. Validation criteria. Ideally the criteria are completely known and fixed. This is the case for example if a protocol is modelled for which the intended service is clear. If this service can be faithfully translated into the validation language all is simple and well. It often occurs, however, that the criteria are made up during validation, cannot be fully expressed in the validation language, or that the validation language is too abstract, too remote from the validator's frame of mind, that errors are made during the translation of the informal requirements.

Note that PROMELA/SPIN does not clearly distinguish between the model and the validation criteria. For example, `asserts`, monitor processes and various labels are added to the model instead of being held separate. This is actually a strong point in many cases, since it allows to express criteria in different ways, suited to the problem at hand.

4.4. Validation. A validator using (X)SPIN is confronted with many choices. First of all, there are the SPIN system parameters such as the maximum memory size, the estimated state space size, the maximum search depth, whether partial-order reduction is used or not, and whether exhaustive or supertrace search mode is used. These parameters mainly determine the performance/coverage of the system. Guidance on how to choose them is provided on one of the help screens of XSPIN.

Next, there are SPIN parameters that determine how many errors are reported and the kind of property checked: safety properties, non-progress cycles, acceptance cycles, or a property expressed as a never claim. These, together with the validator's choices of

1. channel sizes and other constants,
2. which failure mode of the environment is included (semantics of timeout, message loss, node failure, etc.),
3. which specific property or combination of properties is put into the never claim,
4. which scenario or scenarios is exercised, and
5. when the system has multiple layers: the order in which the layers are traversed,

determine a many-dimensional landscape. The validator's itinerary through this landscape has to be chosen judiciously in order to get the most out of validation in the shortest time.

Regarding failure modes, it seems wise to start with a friendly environment (free of errors) and add failures gradually. If the design has to be changed, one may have to check again, to be sure that it also works correctly without the added failure behaviour. Theoretically it is possible that for example progress is only assured thanks to this failure behaviour.

Regarding multiple layers, reduction techniques "outside" the tool (see section 4.2) may be useful when a layered system is validated. One can work from the bottom up. After validating a layer, it may be reduced to a simpler model, cf. the `pftp` example in [**Hol91**]. An alternative is to replace the lower layer by its most general behaviour and put the validated properties as assumptions into the never claim.

4.5. Validation result. When an error is found, SPIN produces a trail that leads to a violating state or through a violating cycle. The problem at this point is to determine the *cause* of the error and fix the model accordingly. Little can be said in general on how to do this.

An interesting instance of this problem arises when checking for a conjunction of several temporal properties. Suppose that one wants to check that a model statisfies two temporal properties `p` and `q`. One could translate each of them into a never claim and validate the model against these claims, one after the other. Alternatively, in an effort to reduce the number of runs, one could combine the two properties into one property `p && q`, translate this into a never claim and validate the model once. The problem now is, that if a violation is found of the combined property, it may not be clear which of the elementary claims is violated.

A seemingly trivial but interesting question is when to stop validating; the same question also arises in the context of testing. The answer depends on the objective and whether the validation criteria are complete and fixed beforehand. If the objective is to show correctness using fixed criteria and no errors are found, validation can be considered done within the limits of the language/tool. If the objective is to get as many errors as possible (as in the BBI case), or the criteria are not fixed, it depends on when one is satisfied with the results and the trade-off between the effort spent on further validation and the gravity of the errors searched for.

Note that validation is not an end in itself but embedded in a larger activity: the production of a correct system implementation. The objective of validation is usually to get confirmation that a system is indeed correct, or ar least to get more confidence in the correctness, since full correctness will in general not be reachable in practice.

5. Suggestions for extensions

Some of our suggestions have already found their way into SPIN. Among these are:

- The use of breakpoints in interactive simulation
- Skipping part of the output on long, guided simulation runs
- Finding the shortest trail to error states.

Other items on our wish list concern extensions of the basic tool, and extensions of the specification language.

Tool

- Support for a mixture of simulation and validation (SPIN).

  Sometimes one wants to validate only part of a larger design. It would be useful to be able to interactively guide the model to a particular part and run a validation from that point onwards. The design of SPIN currently distinguishes between simulation and validation. One could use `gotos` to guide the model to a specific part, but that is not a general solution, since also channels and variables need to acquite their correct initial values. If not done with care, an unreachable state might be the result. Interactive simulation then gives a better solution.

  Another solution would be the ability to take the synchronous product of the model with another special-purpose one that does the steering. A never claim operates in this way, but it may preclude the checking of some properties.
- Mixed simulation.

  Guided simulation is used only for rerunning a trail. You cannot rerun a trail and then start an interactive simulation to explore the state space after the end of trail.
- Enhancement of the user interface (XSPIN).

  Before the introduction into SPIN of an option to find the shortest error trail, one had to do a binary search on the search depth to find the least depth that still produced an error. Now this search is provided at the press of a button. Similar repetitious jobs that are ideally moved from the user to the tool are cycling through the range of properties to be validated, and validating several never claims in sequence.

  A possible solution is to use a meta-language in which validation strategies can be programmed, cf. ML which originated as the "interface language" for the theorem prover LCF. Tcl/Tk, the implementation language of XSPIN, could be used for this purpose.
- A management system around XSPIN would be extremely useful: to register models, validation runs, trails and outputs.
- Display/output of the reachable state space as a Finite State Machine. This is useful to check one's intuition of language features on small models or as an interface to other tools.

- A simple extension of the LTL interface of XSPIN: allowing abbreviations for combinations of temporal operators would be welcome. For example if one could define `p Unless q = (p U q) || []p`, then `p Unless q` would be much more readable than `(p U q) || []p`, particularly when `p` is itself a complicated expression.

Language extensions

- More structuring facilities, e.g.
  - A CSP/LOTOS/OCCAM-like parallel operator.
    The interleaving parallel operator is simple to implement, it can be done with the current language using macros (fork and wait); The synchronized parallel operator is more powerful and will be harder to implement. Of course the effectiveness of PROMELA should not be compromised, so an option is to provide this kind of language extensions by translators.
  - Procedures/functions instead of *cpp* macros.
  - A more refined name-space hierarchy.
    Consider for instance a model of a distributed system with two protocol stacks. Currently it is not possible to separate the global data of one stack from the other, leading to large global data areas.
    A solution might be to allow subprocesses running in the same space as their creator. Syntax: proctypes declared within proctypes (cf. Pascal). Problem: a never claim may only access global data.
- The ability to specify for each channel separately whether it blocks or not.
- Allow a `d_step` within an `atomic` sequence; then `d_steps` can truely be regarded as new primitive PROMELA instructions.
- More control over exiting of processes.
  Processes do not disappear immediately after their last action but live on as zombies. This can be a nuisance when modelling a dynamic network of terminating processes. In particular, Message Sequence Charts quickly become too wide to handle.
- A minor issue, but a synchronous channel used only for synchronization (i.e. without passing data) could be written as `chan c = [0] of { }` and input/output on such a channel by `c?` and `c!`. Instead, PROMELA now requires a dummy data type, e.g. `chan c = [0] of { bit }`, and input/output on such a channel by e.g. `c?0` and `c!0`.

Other items

- A formal semantics definition of PROMELA, e.g. a structured operational semantics. Setting up such a semantics will resolve fine issues such as the interaction between (new) language features.
- Comprehensive documentation, including a tutorial describing all features of PROMELA and SPIN with examples of their use, and a cookbook with examples of modelling idioms.
- Support for (discrete) time.
- A description of the various output formats of SPIN to facilitate extraction of relevant information.

6. Concluding remarks

The use of SPIN and PROMELA has improved the design of the BOS system. There are several ways in which PROMELA/SPIN could be useful later on in the project:

- Support for testing.
  PROMELA may be used as a test specification language. It has many features similar to TTCN, the standardized notation for protocol tests. Test derivation algorithms, in particular for (non-deterministic) finite-state machines or labelled transition systems, could be used to derive PROMELA tests for PROMELA models. Alternatively, test derivation algorithms for languages like LOTOS could be used by translating PROMELA models into LOTOS.
- Script validation.
  The decision algorithm for the control of the barrier is not fixed, but can be changed. A special language has been designed to encode the algorithm. This encoding of the algorithm, called a "script", is then interpreted by the BOS system. Translation of a script to PROMELA seems feasible, making it amenable to validation. A problem will be modelling the environment of the script interpreter, which is basically the whole BOS system.
- Support for getting from design to C++ code.
  Code generation from PROMELA might be useful here.

Finally, we suggest some research subjects not mentioned already:

- Combined model checking and symbolic verification.
  Especially where models are parameterized, e.g. by channel capacity or by configuration tables, symbolic verification could be used to show that it is sufficient to validate a particular finite model.
  A related issue: include algorithms for checking boundedness of channels, and lift algorithms for establishing deadlock-freedom based on the communication pattern of processes to PROMELA (such algorithms exist e.g. for CSP and occam).
- Cheapest way to express a property to be validated.
  There are several ways to express correctness criteria in PROMELA; a theory that relates these criteria and allows to select the "cheapest" could be useful.
- Translation of various languages to PROMELA, to make model checking available for those languages.
  One could think of LOTOS or Occam as candidates. Occam is another descendant of CSP('78) and hence shares many features with PROMELA. Notable differences: occam has `PAR` instead of `run`, procedures/functions (even locally declared ones), timers, priorities on guards, guards containing both a Boolean expression and an I/O guard, and variable length arrays as channel data. Moreover, occam 3 has modules and union types.
  The major problem will be to find translations that produce small models.
- Investigation of the relation between the π-calculus and PROMELA (note that channels are treated as just another data type).

Acknowledgements. Thanks to the BOS project team at CMG, The Hague, in particular the chief designer Klaas Wijbrans and Eric Burgers, who did much of the work. Thanks to Gerard Holzmann for his instant reaction to our questions and solutions to our problems. Thanks to the other members of the "BOS team" at the University of Twente: Ed Brinksma, Wil Janssen and Job Zwiers, and to my colleague Rom Langerak for discussions on PROMELA/SPIN.

References

[BF91] Ricky W. Butler and George B. Finelli, *The infeasibility of quantifying the reliability of life-critical real-time software*, Software Engineering Notes **16** (1991), no. 5, 66–76.

[Cap93] Cap Volmac (Cap Gemini Sogeti Group), Utrecht, The Netherlands, *SDW — System Development Workbench*, 1993.

[Erd95] Hakan Erdogmus, *Verifying semantic relations in Spin*, (Jean-Charles Grégoire, ed.), URL `http://netlib.att.com/netlib/spin/ws95/spin95_abstracts.html`. 1995.

[Hol91] Gerard J. Holzmann, *Design and validation of computer protocols*, Prentice-Hall, 1991.

[Spi92] J. M. Spivey, *The Z notation: A reference manual*, 2nd ed., Prentice Hall, 1992.

[TC96] Stavros Tripakis and Costas Courcoubetis, *Extending Promela and Spin for real time*, Tools and Algorithms for the Construction and Analysis of Systems, Proceedings of TACAS'96 (Tiziana Margaria and Bernhard Steffen, eds.), 1996, LNCS 1055, Springer, pp. 329–348.

[Wij96] Klaas Wijbrans, *The BOS project: practical experiences with formal specification*, Proceedings of the Third Dutch Specification Day (Rom Langerak, ed.), 1996.

TELE-INFORMATICS AND OPEN SYSTEMS GROUP, DEPARTMENT OF COMPUTER SCIENCE, UNIVERSITY OF TWENTE, P.O. BOX 217, 7500 AE ENSCHEDE, THE NETHERLANDS.

E-mail address: `kars@cs.utwente.nl`

DIMACS Series in Discrete Mathematics
and Theoretical Computer Science
Volume **32**, 1997

Implementing and Verifying MSC Specifications Using PROMELA/XSPIN

Stefan Leue and Peter B. Ladkin

ABSTRACT. We discuss a translation of Message Sequence Charts (MSCs) into the language PROMELA (we call this translation an '*implementation*') that is consistent with the formal semantics we have previously defined for Message Flow Graphs and Message Sequence Charts, which handled the syntactic features with mathematical import from ITU-T recommendation Z.120. We report on experiments executing the PROMELA code using the XSPIN simulator and validator. In previous work we found that potential process *divergence* and *non-local choice* situations impose problems on implementations of MSCs, and we discuss how these impact our PROMELA translation and suggest solutions. Finally, we show how to model-check liveness requirements imposed on MSC specifications. We use the PROMELA models obtained from our implementation, describe how to use control state propositions based on these models, use Linear Time Temporal Logic formulas to specify the liveness properties, and demonstrate the use of XSPIN as a model checker for these properties.

1. Introduction

Message Sequence Charts (MSCs) describe sequences of message exchanges by communicating, concurrent processes. While other specification languages like SDL or PROMELA describe the process behaviour explicitly, leaving message flows to be inferred, MSCs specify explicit message flows while other details of process behaviour must be inferred from the specification. The syntax of MSCs is described in the ITU-T Recommendation Z.120 [**IT96**]. MSCs are frequently used both formally and informally for the description of message flow amongst communicating, concurrent processes. They have found their way into many software engineering methodologies and toolsets, such as SDL tools and environments [**OFMP$^+$94, Lab95**], Object-oriented methodologies [**RBP$^+$91, Jea92, SGW94**], tools to analyse the design of message exchanges at early stages in the software lifecycle [**Hol96, AHP96**], and methods describing design patterns [**BMR$^+$96**].

In previous work we defined a finite state semantics for Message Sequence Charts [**LL95b**], and discussed in [**LL95a**] implications of the MSC notation as

1991 *Mathematics Subject Classification.* Primary 68Q60, 68Q55, 68Q10; Secondary 68N99, 03B45.

The first author was supported in part by the National Science and Engineering Research Council of Canada (NSERC).

defined in Z.120. Although Basic MSCs look simple, syntactic features such as *conditions* or branchings which are defined for *High-level MSCs* can make it harder to figure out what these latter describe and automated help is desirable. We describe a translation of MSC specifications into PROMELA so that they can be simulated and validated using the XSPIN tool. We call this translation an *implementation*, and discuss in this paper what design choices need to be made.

There are three major ways in which MSCs are used:

1. to visualise actual system execution, during debugging and program understanding (as in [**Hol96, AHP96**]),
2. as a language to document early design decisions (as in ObjecTime [**SGW94**], EFD [**Hol96**]),
3. to document test cases or functional-validation criteria that an implementation must satisfy (ROOM [**SGW94**], SDT [**AB**]).

Items 1 and 3 concern finite execution scenarios in which one event node corresponds to a single event and a single message. Z.120 calls such finite MSCs *Basic MSCs* (BMSCs). For an example of a BMSC, see the left-hand side of Figure 1. The intuitive meaning of this BMSC is as follows. There are two processes `P1`, `P2` with vertical 'time lines', bounded above and below. `P1` sends a message of type `a` to `P2`, which (asynchronously) receives it, and then sends a message of type `b` to `P1`, which (asynchronously) receives it. This example should suffice to understand what BMSCs are supposed to mean. BMSCs do not have any branching or iteration. In [**LL95b**] we described how to represent BMSCs algebraically as so-called basic *Message Flow Graphs* (MFGs), and then translated MFGs into (global-) state machines. We are not primarily interested in BMSCs here since their meaning is straightforward and implementation is trivial. We concern ourselves with MSCs describing repeating or infinite behavior, in which a given event 'node' typically represents many repeating events in an execution sequence (as, for example, statements in loops in procedural programming languages).

Composition of MSCs. When MSCs are used at early system design stages to represent desired behaviour of the system, many BMSCs will typically be written, and these sets of BMSCs only make sense if some sort of relationship between individual BMSCs is intended. In practice, a single BMSC often corresponds to a particular software feature, described by a finite message exchange *scenario*. In Figure 12, for example, the MSC labelled `MSC1` represents the scenario in which process `P1` is requesting connection establishment from process `P2` by sending a `CR` message (*connect request*), while `MSC2` shows `P2` answering by a `CC` message (*connect confirm*) and `MSC3` shows `P2` answering by a `DR` message (*disconnect request*). Intuitively, these scenarios form the building blocks for a simple connection-establishment protocol, provided their relationship is properly defined.

Explicitly to represent this relationship formally calls for some sort of composition operator. The latest version of the Z.120 standard introduces so-called *High-level MSCs* (HMSCs) to specify the interrelation of MSCs[1]. HMSCs may represent branching and iterating behaviour. Composition is described by a graph that we will call an *HMSC graph*. We use the following definition of HMSC. An HMSC

[1]According to Z.120, both *conditions* in BMSCs and *composition* expressed by HMSCs can be used simultaneously. The expressive capability of HMSCs is greater because it allows expression of the n-fold repetition of a BMSC M for a fixed, finite n.

graph is a graph with *start nodes* (nodes with only out-edges), *end nodes* (nodes with only in-edges) and *interior nodes* (nodes with both out- and in-edges). Each interior node is labelled either with a BMSC or with another HMSC graph. We assume that there is at least one start node and one interior node. The intuitive meaning is that the edges of an HMSC graph indicate control flow between BMSCs (or other HMSCs) that are the node labels. The composition of two BMSCs is thus represented by an edge between the corresponding nodes in the HMSC graph[2].

We define an *MSC specification* to be a collection S of one or more BMSCs, plus an HMSC graph G whose nodes are labelled with members of S. The MSC Specification in Figure 1 specifies a finite scenario in which process `P1` sends an `a` message to `P2` which replies by sending a `b` message. The MSC Specification in Figure 12 specifies branching and iterating behaviour. It can be interpreted as specifying a simple connection establishment protocol in the following way: the request for connection establishment (node `MSC1` in the HMSC on the right hand side of Figure 12) can be followed either by connection confirmation (node `MSC2`) which means that the protocol ends, or by request of disconnection (node `MSC3`) which means that a new connection establishment must be attempted (loop back to node `MSC1`).

HMSC graphs hold out the hope for a clearer notion of MSC composition than possible with *conditions*. We are particularly interested in MSC specifications which involve HMSCs containing cycles: it is intuitively only possible to 'visit' a given MSC node (corresponding to a communication event in the MSC) more than once in an execution sequence if there is some control path leaving that node which returns to it; the control path is some (here unspecified) construct of the BMSC of which that node is part plus edges in the HMSC graph.

Many of the problems in interpreting iteration and branching in MSCs noted in [**LL95b, LL95a**] occur regardless of the syntactic form in which a composition is proposed. The current Z.120 HMSC proposal does not appear to contain syntactic restrictions that would avoid many of these interpretation problems. We will show that they become apparent when pursuing PROMELA simulations of MSC specification.

Motivation. Our reasons for implementing (i.e., simulating) MSCs are:

1. we want to demonstrate the practical use of MSCs in behavior specification; in particular,
2. we want to demonstrate the practical use of our semantics;
3. the synthesis of process code from MSCs requires an understanding of what behavior they express and the assumptions they embody, which can be most easily seen from simulation;
4. this exercise in translating MSCs into process code reveals underspecified assumptions in the intended meaning of the MSCs;
5. we want to generate models that can be used for model checking properties of MSC specifications.

Choice of PROMELA/XSPIN. We chose PROMELA/XSPIN because PROMELA provides all necessary concepts (sending and receiving primitives; parallel and

[2]An HMSC graph that intuitively represents a single BMSC may be constructed as follows: a single start node leads to a single interior node labelled with the BMSC, leading to a single end node. Thus these particular HMSC graphs can be identified intuitively with the BMSCs with which they are labelled.

asynchronous composition of concurrent processes; and communication channels) that were necessary to implement MSC specifications. Furthermore, the XSpin [**Hol91, Hol**] tool allows for randomly simulating Promela specifications, which helps in debugging, and for model-checking properties expressed as LTL formulas. The availability of suitable language features and the simulation capability distinguishes Promela/XSpin from other finite-state modelling-language and model-checker packages such as SMV [**McM93**]. The communication primitives and channels that are readily available in Promela would need to be hand-coded into SMV specifications in order to obtain models identical to the ones we obtain from our Promela implementation.

2. A few observations

Our MSC formal semantics [**LL95b**] allows us to make the following observations:

- MSCs describe asynchronous communication, as in Z.120. However, our semantics also deals with synchronous communication (see [**LL95b**]). Thus so does the Promela translation. We restrict ourselves to discussing the asynchronous case here.
- MSCs are inherently finite-state. The state of an MSC specification is determined by the state of each process (i.e., at which point the process control lies in each process), and by the "state" of each of the messages in the system (whether the message is on the way, or not). Since there are only finitely many processes, and each process has finitely many control states, and there are a finite number of message arrows in an MSC specification, and these are all the meaningful 'parts' of an MSC, there is a strong *prima facie* argument that there are only finitely many states. For additional arguments for a finite-state interpretation, see [**LL95a**].
- Some claim that the communication between processes in MSCs is buffered. If so, the behavior of the buffers is completely hidden. This could lead to trouble – generally, specifications should be explicit about everything they deal with. We feel that for a specification style based on graphics, *what you see* should be *what you get*. If not, *what you get* are *problems*.
- Liveness properties in MSCs are underspecified. See [**LL95b**].
- Even somewhat restricted use of basic MSC composition yields specifications with problematic meaning. Sense may be made of them only if substantial assumptions are made about the behavior of the environment. Difficult cases arise from message cross-over (as in Figures 3 and 7), as well as non-local choice points as in Figure 14. See [**LL95a**].

3. Implementation of Basic MSC specifications.

3.1. Message Flow Graphs. MSC specifications are graphical objects - ink on paper, lines on a screen. To implement an MSC, one must translate the graphical into a textual or mathematical representation. Z.120 proposes a textual syntax. We pointed out ambiguities in the mapping from the graphic to this textual representation in [**Leu94**]. The Z.120 textual syntax is therefore not suitable for our purposes.

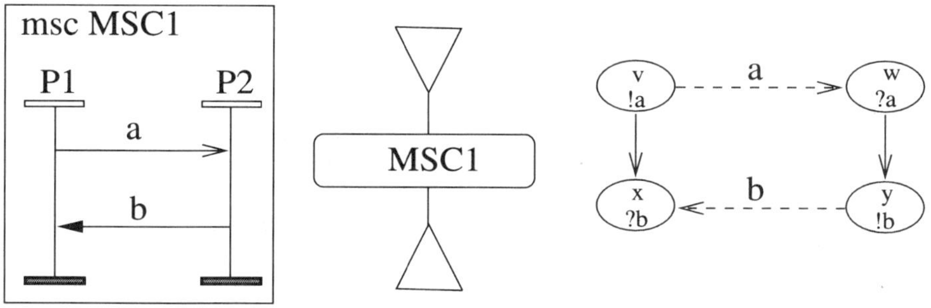

FIGURE 1. MSC example 1

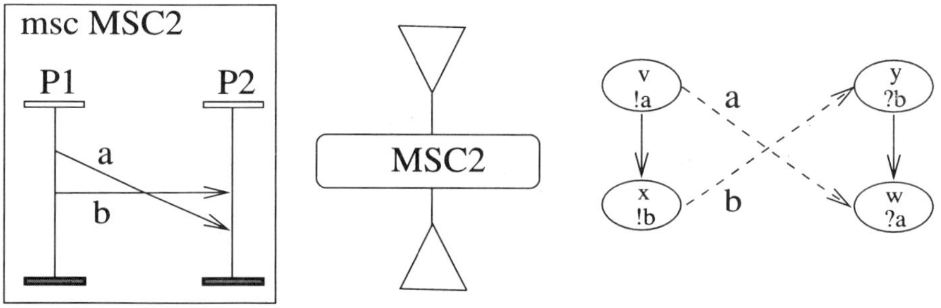

FIGURE 2. MSC example 2

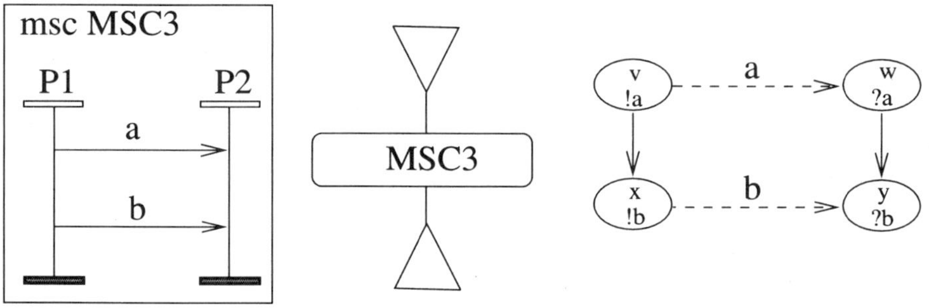

FIGURE 3. MSC example 3

In [**LL95b**] we defined the mapping of MSCs to Message Flow Graphs (MFGs), where each node in the MFG corresponds to a communication event in the corresponding MSC. An MFG corresponding to the MSC example 1 is given on the right hand side of Figure 1. The nodes are connected by directed arrows representing two relations on the set of nodes: the *next-event* (*ne*) relation representing the control flow in a process, and the *signal* (*sig*) relation representing message flows. In this paper, we draw *ne*-relation arrows as solid lines and *sig*-relation arrows as

```
mtype = {a, b};

chan vw = [1] of { byte };
chan xy = [1] of { byte };

proctype P1()
{ atomic {
     vw!a;
     printf("!a\n") };
  atomic {
     xy?[b] -> xy?b;
     printf("?b\n")}
}

proctype P2()
{ atomic {
     vw?[a] -> vw?a;
     printf("?a\n")};
 atomic {
     xy!b;
     printf("!b\n")}
}

init { atomic { run P1(); run P2() } }
```

FIGURE 4. PROMELA code for MSC example 1

dashed lines. The translation of an MSC specification into a corresponding MFG is described in [**LL95b**]. While there the concept of *conditions* was used to specify possible continuations of one MSC by another, in this paper we use the HMSC concept, recently introduced into Z.120, to specify composition of basic MSCs. The translation procedure from MSC specifications into MFGs based on HMSCs is a straightforward extension of the *"unfolding"* operation as defined in [**LL95b**], and we will not elaborate on this translation here.

We distinguish two types of MSC specifications:

1. MSC specificatios may describe *finite* system executions. This is depicted in Figure 1. In this case the corresponding MFG is a finite, cycle- and branching-free graph.
2. MSC specifications may represent *branching* and *iterating* behaviour, which is reflected by the branching and cyclic structure of the corresponding MFG (see the MFG in Figure 13 that corresponds to the MSC specification in Figure 12).

Note that neither the HMSC concept nor the condition concept in Z.120 imply any sort of synchronization between processes in an MSC specification when sequential composition or branching occurs[3]. As we shall see later, this has bearing on our implementation choices.

MFGs will be the basic underlying data structure for the implementation although we will in most cases not explicitly refer to them. In the remainder of this section we discuss the implementation of basic MSC specifications. In Section 4 we will discuss the implementation of iterating and branching specifications.

---

[3]C.f. [**IT96**]: "*A sequential execution of two nodes that are related by an edge is described byt the* seq *operator.*" and "*The* seq *operator denotes the weak sequencing operation where only events on the same instance are ordered*". Concerning the semanitcs of conditions, c.f. [**IT95**]: "*Note that the semantics of a chart containing conditions is simply the semantics of the chart with the conditions deleted from it.*"

3.2. Basic implementation concepts. We model an MSC specification in PROMELA by instantiating a PROMELA process for each of the MSC processes at system-setup time[4]. This is implemented by an `init` clause in PROMELA. For the concurrent instantiation of two or more processes, we need to employ the `atomic` keyword. For example, to initialize an MSC with two processes `P1` and P2 we write `init { atomic { run P1(); run P2() } }`. Messages can have types in PROMELA as well as in MSCs. We choose the `mtype` construct to specify the message types. `mtype = {a, b}` generates two one-byte integer constants with names `a` and `b` and increasing values `a`=1 and `b`=2.

To model the message behaviour of MSCs in PROMELA we choose channels with capacity 1, one for each message arrow in the chart. This represents the invariant that any given message is either on the way (in which case there is a message of the expected type in the PROMELA channel), or not on the way [**LL95b**]. The channels must have type consistent with the message type. In PROMELA, channels are implemented as arrays of finite length ≥ 0. The declaration `chan vw = [1] of { byte }` defines a channel with name `vw` and capacity of one element of the message type – in this case, one byte.

A PROMELA implementation of an MSC has the following overall syntactic structure (c.f. Figure 5):

- First, necessary data definitions, including the global channel declarations denoted by the keyword `chan`.
- Next, the definition of the process bodies as indicated by the keyword `proctype`. In our examples the processes do not have parameters – all names used (i.e., the channels) have global scope.
- Finally, the instantiation of the whole system using an `init` statement.

3.3. Basic MSC specifications. *MSC example 1.* The left hand side of Figure 1 shows a basic MSC in which process `P1` sends a message of type `a` to `P2` and `P2` then sends a message of type `b` back to `P1`. Figure 4 shows the code for the MSC in Figure 1. We have defined two processes, as in the MSC. The core of each process implements the communication behaviour, plus instructions to print output to the screen for debugging purposes (the `printf` statements). The statement `vw!a` denotes a *send* of a message of type `a` over the channel named `vw`, and `xy?b` denotes reception of a message of type `b` from channel `xy`.

The semantics in [**LL95b**] relies on the interleaving model with communication events as atomic actions[5]. PROMELA requires the use of the `atomic` keyword to ensure that operations inside the following curly parentheses are executed as an atomic action, without other interleaved events. We therefore ensure that the execution of the communication statements and the related `printf` debugging statements are atomic events by use of the `atomic` keyword.

Reception of messages in PROMELA is not blocking. Thus, when executing an `xy?b` statement, a message of type b will be received if there is a message of that type at the head of the channel `xy`. However, if there is no such message at the head, the statement will nevertheless be executed, a message of undefined type will

[4]Note that a 'process' is a connected component of the *ne* relation of an MFG [**LL95b**]

[5][**Sel96**] chooses a receive event and a subsequent send event to be executed in one atomic transition of a ROOM actor. However, they derive code for one single test actor from an MSC specification. Interleaving semantics, however, is only relevant for systems consisting of more than one concurrent processes.

```
mtype = {a, b};

chan vw = [1] of { byte };
chan xy = [1] of { byte };

proctype P1()
{ atomic {
    vw!a;
    printf("!a\n") };
  atomic {
    xy!b;
    printf("!b\n") }
}
proctype P2()
{ atomic {
     xy?[b] -> xy?b;
     printf("?b\n")};
  atomic {
     vw?[a] -> vw?a;
     printf("?a\n")}
}
init { atomic { run P1(); run P2() } }
```

FIGURE 5. PROMELA code for MSC example 2

```
mtype = {a, b};

chan vw = [1] of { byte };
chan xy = [1] of { byte };

proctype P1()
{ atomic {
    vw!a;
    printf("!a\n") };
  atomic {
    xy!b;
    printf("!b\n") }
}
proctype P2()
{ atomic {
     vw?[a] -> vw?a;
     printf("?a\n")};
  atomic {
     xy?[b] -> xy?b;
     printf("?b\n")}
}
init { atomic { run P1(); run P2() } }
```

FIGURE 6. PROMELA code for MSC example 3

be received, and process control will advance beyond the reception statement. This doesn't happen in MSCs, which block on *receive* of a message that isn't there.

In order to implement blocking on reception we use a guard, namely a predicate which checks whether a message of the suitable type is ready to be received. This is the `xy?[b]` statement, which is true if the first element of the channel `xy` is of type `b`, and false otherwise. The `->` operand serves as an enabling operator such that the operation on its right is only enabled if the guard on its left is *true*. In order to protect the compound guarded receive statement from undesired interleaving, it needs to be embraced by an `atomic` statement. The execution of this example using SPIN yields exactly one execution trace.

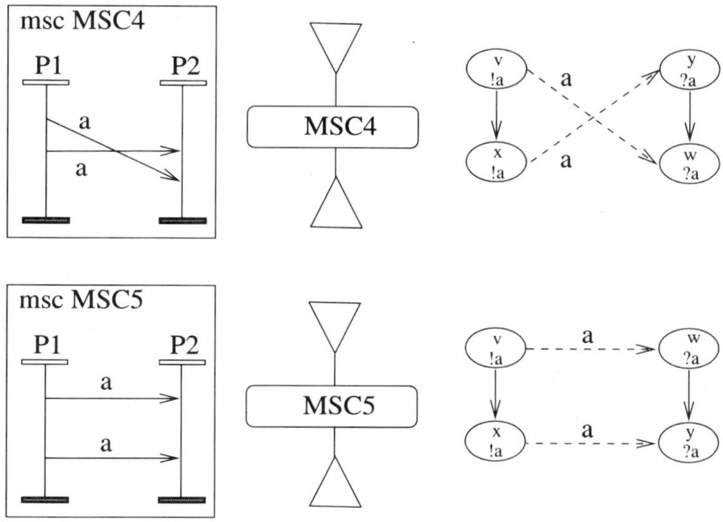

FIGURE 7. MSC example 4 (top) and 5 (bottom)

MSC example 2. The MSC example 2, Figure 2, is similar to the MSC example 1. However, we inverted the direction of the message arrow of type b. This MSC specifies a "message overtaking" – message b is sent later but received earlier than message a. This gives rise to the use of per-message dedicated channels in PROMELA because message-overtaking within a PROMELA channel is not possible. As for MSC example 1, execution using SPIN yields exactly one execution trace.

MSC example 3. Example 3 allows two execution sequences: <!a, !b, ?a, ?b> or <!a, ?a, !b, ?b>. After the !a event has occurred, two independent events are enabled: the !b and the ?a event. The PROMELA semantics specifies a nondeterministic choice in this situation. SPIN implements the nondeterministic choice in PROMELA by using a random (non-repeating) choice. As expected, experimentation with the SPIN simulator shows that two different traces will be generated (see also [**LL96**]).

MSC examples 4 and 5. Examples 4 and 5 in Figure 7 are similar to Examples 2 and 3, except that both message arrows are of the same type (a). The PROMELA implementation generates similar outputs to those in examples 2 and 3 except for !b replaced by !a and ?b replaced by ?a. Example 4 generates <!a, !a, ?a, ?a> as trace, whereas example 5 generates <!a, !a, ?a, ?a> or <!a, ?a, !a, ?a> as traces, randomly chosing between them. In [**LL95a**] we discussed an anomaly arising with these two examples. Both specifications stand for the same "code" with respect to the communication events, namely for two consecutive statements of type !a for the left and of type ?a for the right process. In other words, the left and the right processes in both examples are code-identical. However, they do not allow the same set of traces. We conclude that there must be an implicit assumption about the environment which distinguishes the specifications. What would this be?

```
mtype = {a};
chan vw = [1] of { byte };
chan xy = [1] of { byte };

proctype P1()
{ atomic {
vw!a;
printf("!a, 1\n") };
atomic {
xy!a;
printf("!a, 2\n") }
}
proctype P2()
{ atomic {
vw?[a] -> vw?a;
printf("?a, 1\n")};
atomic {
xy?[a] -> xy?a;
printf("?a, 2\n")}
}
init { atomic { run P1(); run P2() } }
```

FIGURE 8. PROMELA code for ex. 4

```
mtype = {a};
chan vw = [1] of { byte };
chan xy = [1] of { byte };

proctype P1()
{ atomic {
    vw!a;
    printf("!a, 1\n") };
  atomic {
    xy!a;
    printf("!a, 2\n") }
}
proctype P2()
{ atomic {
     xy?[a] -> xy?a;
     printf("?a, 2\n")};
  atomic {
     vw?[a] -> vw?a;
     printf("?a, 1\n")}
}
init { atomic { run P1(); run P2() } }
```

FIGURE 9. PROMELA code for ex. 5

The implementations given here support this conclusion. In order properly to implement the desired behaviour we needed to define that both messages are implemented by different channels. The channels belong to the environment. The processes thus receive the messages from distinct environment entities, which allows for modeling the faster delivery of one message than of the other. This indicates that in an implementation we indeed need to exploit the environment to get the "expected" behavior.

4. Iterating and branching MSC specifications

As discussed above, HMSCs may specify a composition of basic MSCs such that an iterating or branching system is described; iterating and branching MSC specifications translate into iterating and branching MFGs (Figure 10).

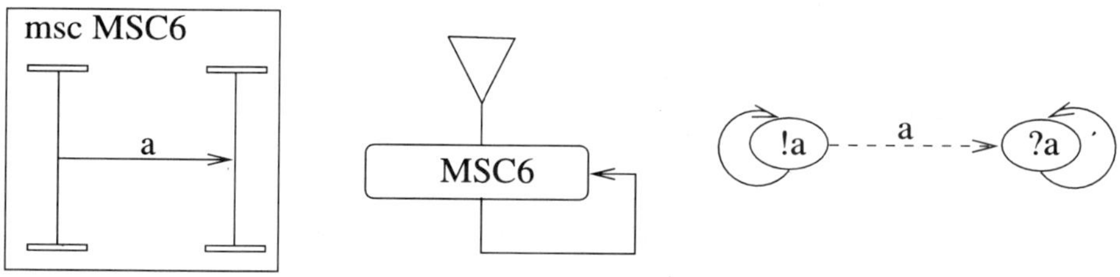

FIGURE 10. MSC example 6

```
mtype = {a};
chan vw = [1] of { byte };
chan xy = [1] of { byte };

proctype P1()
{ C:
  atomic {
    if
    :: vw!a
    :: full(vw) ->  printf("full,  ")
    fi;
    printf("!a, "); goto C}
}
proctype P2()
{ C:
  atomic {
    vw?[a] -> vw?a; printf("?a, "); goto C}
}
init { atomic { run P1(); run P2() } }
```

FIGURE 11. PROMELA code for MSC example 6

4.1. Iteration. Figure 10 shows an HMSC which allows `MSC6` to be followed by `MSC6`. In PROMELA, we model the iteration in the process code by a `goto`–label construct. We must consider one aspect of the semantics of PROMELA more thoroughly. We see that a process may reach a *send* statement repeatedly. The sending primitive in PROMELA is blocking, i.e. when the channel `vw` is full, the statement `vw!a` blocks. MSCs according to Z.120, on the other hand, do not have the notions of channels and capacities. It would therefore be counterintuitive if an MSC could block on sending. This means that we need to add a void operation which is carried out if the send operation blocks. We use the PROMELA predicate `full(vw)` as a guard which is true if and only if the actual send operation `vw!a` blocks. We use the dummy `vw?[a]; printf("full, ")` statement to indicate that a write operation to a full channel was attempted.

In the following prefix of a (supposedly) infinite execution trace of this example note that in conformance with our semantics the execution of one receive statement `?a` may disable $n \geq 1$ send operations `!a`.

```
swen12:/swen12/u/sleue/spin/specs/workshop.366 % spin msc6a.prm
!a, ?a, !a, full, !a, ?a, !a, full, !a, ?a, !a, full, !a, full, !a,
?a, !a, ?a, !a, ?a, !a, ?a, !a, ?a, !a, full, !a, full, !a, ?a, !a,
```

It is worth noting that there is nothing in the MSC specification which would make an infinite sequence of `!a` events an illegal trace. In other words, there is nothing in the MSC specification which would ever require a `?a` event to occur [**LL95b**].

However, the algorithm which resolves nondeterminism in PROMELA (based on a random number generator, we believe) appears to ensure some fairness condition on the nondeterministic choice alternatives.

4.2. Branching. The MSC example in Figure 12 specifies *branching* behaviour. It may be interpreted to specify a very lightweight connection establishment protocol: process `P1` requests establishment of a connections by means of a `CR` protocol data unit (PDU); depending on a non-deterministic decision, `P2` either acknowledges the establishment by a `CC` PDU, or refuses connection establishment by a `DR` PDU. In the first case the system execution is assumed to go into a data-transfer phase (here indicated by a triangular symbol indicating an HMSC end-node), in the latter case the system returns to a state from which connection establishment can be re-initiated.

In [**LL95b**] we suggested an operation called *unfolding* to translate an MSC specification into a branching and iterating MFG. We'll adopt that construction here. In order to adapt the unfolding operation for an HMSC, we shall assume that each BMSC in the HMSC has the same processes (i.e., the same number of processes with the same process labels). The unfolding is a single graph that is intended to reflect exactly the behavior intended by the HMSC. For example, when `P1` in Figure 12 has sent the `CR` message, it will have to decide whether to move left (i.e. to continue with the MSC labeled "`MSC2`"), or to move right. The right process is expected to make the same decision after receiving `CR`. Intuitively, the choice of each processes of which 'branch' to follow at the decision point must correspond with the choice (to be) made by the other process at the same point, during the same iteration. These choices can differ from iteration to iteration, as long as the processes make corresponding choices.

Our semantics specifies a non-deterministic decision for both processes, but it does not specify how to implement the decision-making. Let's assume the following strategy: `P2` makes a random decision whether to send `CC` or `DR`. We call this the *random-choice* strategy. `P1`, we assume, remains in its post-`CR` state in `MSC1`, before the 'choice point', monitoring the incoming messages. Depending on whether it sees a `CC` or a `DR` it will react accordingly and continue with either moving through `MSC2` to the 'triangle' or through `MSC3` and back to the beginning of `MSC1`. Let's call this the *wait-and-see* strategy. Obviously, for a consistent implementation of the branching there has to be one process that performs a random choice between sending different messages, and all other processes follow suit by implementing a wait-and-see strategy. We will later see MSC specifications for which such a consistent implementation is impossible.

In the corresponding PROMELA example in Figure 20, the *random-choice* strategy is implemented by a `do ... od` construct embracing two vacuously enabled guarded commands, namely `vw!CC` and `xy!DR`. We rely on the randomness of the choice between both to be implemented by spin. The *wait-and-see* is implemented by a `do ... od` construct which embraces two complementarily enabled statements, namely `vw?CC` and `xy?DR`. In other words, `P1` follows faithfully the decision made by `P2`.

Executing the example with branching control. The following execution traces show that the system runs in iterations until `P2` decides to send a `CC` which leads the system into a terminating state. Again, there is nothing in the MSC specification which would keep it from repeating a `CC` – `DR` loop forever. (We used a variant of

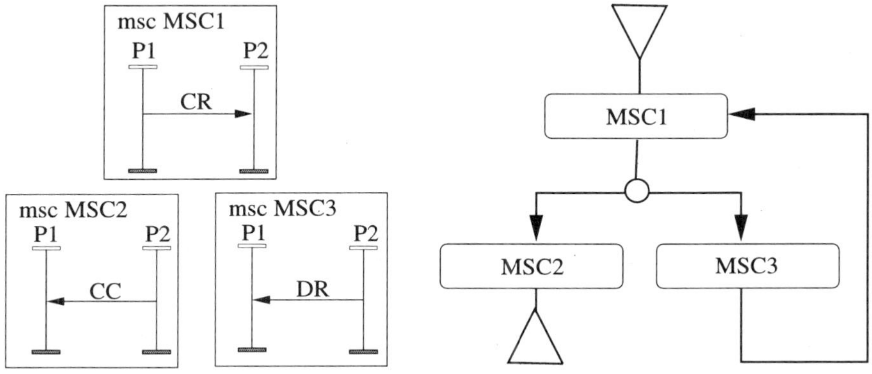

FIGURE 12. MSC Specification example 7

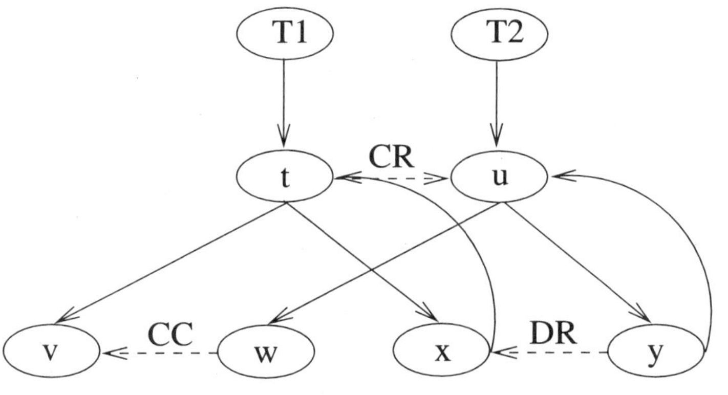

FIGURE 13. Message Flow Graph for example 7

the code in Figure 20 with debugging **printf** statements to generate the following output).

```
swen12:/swen12/u/sleue/spin/specs/workshop.406 % spin msc9.prm
!CR, ?CR, !DR, ?DR, !CR, ?CR, !CC, ?CC, 3 processes created
swen12:/swen12/u/sleue/spin/specs/workshop.408 % spin msc9.prm
!CR, ?CR, !DR, ?DR, !CR, ?CR, !DR, ?DR, !CR, ?CR, !CC, ?CC, 3 processes created
```

4.3. Summary of implementation choices. We summarise the implementation decisions discussed so far. First, the graphical-object MSC specification is translated into a corresponding *Message Flow Graph*, using unfolding. Then:

- Every process (Z.120 terminology: 'instance') in an MSC specification is mapped to exactly one PROMELA process. The PROMELA processes are instantiated concurrently when the whole PROMELA specification becomes incarnated, see the `{ atomic { run P1(); run P2() } }` statement in Figure 20.
- Message arrows are represented by PROMELA channels. This allowed for modeling so-called "message crossing". Also, as messages in MSCs can be

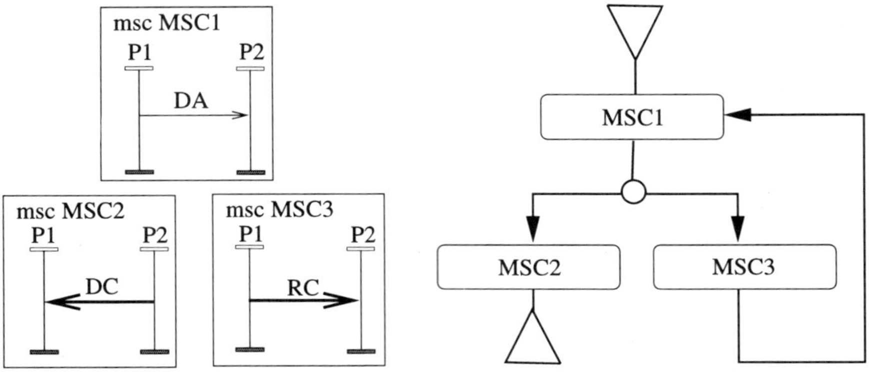

FIGURE 14. MSC specification with non-local choice

exclusively either on-the-way or not, the capacity of these channels is defined to be 1.

- Message *send* and *receive* events are modeled by the corresponding PROMELA statements (e.g., `tu!CR` and `tu?CR`, respectively).
- Branching in MSCs is modeled using PROMELA labels and `goto` statements.
- It is necessary to ensure that certain sequences of PROMELA statements were executed atomically using the `atomic` clause.
- There is no notion of 'channel' explicit in MSCs, therefore there can be no blocking-send statement in the corresponding PROMELA code. We use `full(..) -> skip` statements to model the non-blocking send of an MSC.
- Receive statements, however, are blocking. This is implemented using a PROMELA guard-statement pair, see for an example the `vw?[CC] -> vw?CC` statement pair in Figure 20. It is particularly important to guarantee atomicity of these guard-statement pairs.

5. Implementing Non-Local Choice

The MSC in Figure 14 is similar to the example in Figures 12 and 13. It describes a simple data exchange protocol. `P1` transmits data by a `DA` PDU. Then, two scenarios are possible: either `P2` confirms receipt with a `DC` PDU or, due to some unspecified internal decision, `P1` requests explicit acknowledgement from `P2` through an `RC` PDU.

The implementation of this MSC in PROMELA (Figure 16) illustrates some of the intricacies of using HMSCs (or 'conditions') to compose BMSCs to form MSC specifications: the "*n-th-cycle-same-choice*" condition seems to be what users intuitively understand this MSC to express. This condition says that when `P1` has gone through n iterations of the cycle described by the MSCs `MSC1` and `MSC2`, and if `P1` is ahead of `P2`, then later when `P2` reaches the n-th cycle it will make the same left-right decision in a post-`DA` state that `P1` made in its n-th cycle. If in an implementation `P1` were to decide to "go left" (continue with MSC `MSC2`), and `P2` were to decide to go right (`MSC3`) at the same choice point on the same iteration, then the system would block (it would either dead- or livelock depending on the implementation) because both processes would be waiting for a signal to

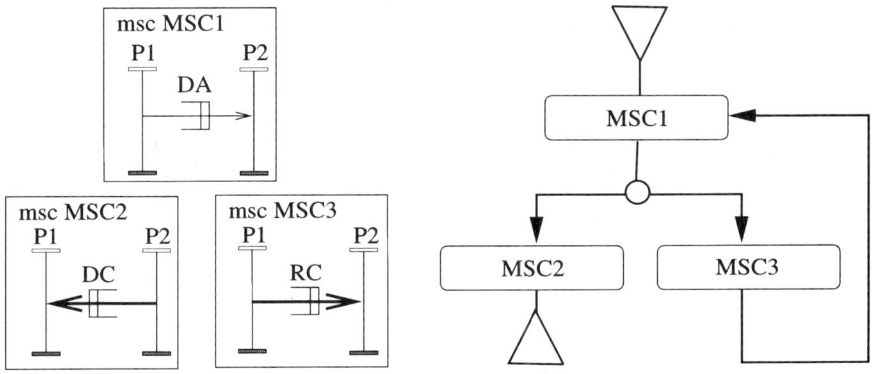

FIGURE 15. MSC example with non-local choice and explicitly represented channels

be sent by the other process. This is not what users understand this MSC to express. Such an "*n-th-cycle-same-choice*" branching was called a 'non-local choice' in [**LL95b, LL95a**] because these choices somehow have to be synchronised by both processes despite the fact they occur at different points in the execution sequence.

It turns out that the synchronisation required by this kind of MSC cannot be implemented in a local, non-coordinated, fashion, in contrast to the situation for MSCs in which a process may branch control without synchronising with other choices, as for example in Figure 12. [**LL95c**] discussed two somewhat unsatisfactory variants of the implementation of the example in Figure 14.

Executing Non-local Choice with history variables. As described in [**LL95a**], existence of a history variable that records the left-right choices is implied by the intuitive meaning of the choice-synchronizing processes `P1` and `P2`. We argued that the length of this variable is finite but potentially unbounded. As PROMELA only allows for the description of finite-state systems, we must bound the size of the history variable, thereby only approximating the history variable algorithm informally described in [**LL95a**].

We use $N+1$ global history variables: variables `i1` ... `i`N and variable **`hist`**. In our example, $N = 2$. Process `P`k keeps track of the iteration it is on by setting history variable `i`k. However, the choice history only records M previous choices (M is thus the bound on the size of the choice-history variable), so `i`k is approximated by `n`k = `i`k mod M. Let $0..N = \{x \mid x \in \mathit{Naturals} \;\&\; 0 \leq x \leq N\}$. The choice-history variable is `hist`: $0..(M-1) \rightarrow \{0,1\}$, initialised to `0`. (There is only one choice-point per iteration in this example and only a two-way choice – **`hist`** would generally be a doubly-indexed array `hist`: $0..(M-1) \times$ *choice-point-labels* $\rightarrow$ *branch-labels*.)

A value `hist[`k`]` $= 0$ indicates that the process first reaching this branch point in the k'th cycle went 'left'; a value of 1 indicates that it went 'right'. (Since the initial value of *hist* also has a meaning as a branch choice, `P`j checks whether `hist[`k`]` has been previously set by checking whether some other `n`k is greater than `n`j, which must be done in any case, as we see next.) `P`k is the only process

```
/* Length of history variable */
#define M 8
/* Channel capacities */
#define Ctu 3
#define Cvw 1
#define Cxy 3

mtype = {DA, DC, RC};
chan tu = [Ctu] of { byte };
chan vw = [Cvw] of { byte };
chan xy = [Cxy] of { byte };

int hist[M];  /* history variable */
int n1;       /* # of iterations through non-local choice by P1 */
int n2;       /* # of iterations through non-local choice by P1 */
int i1;       /* act index to hist for P1 */
int i2;       /* act index to hist for P2 */

proctype P1()
{ C1:
    tu!DA;
  atomic{
    if
    :: (n1 < n2) && (n2 - n1) <= M ->    /* P1 lacks behind */
       if                                /* Decide according to hist */
       :: hist[i1] == 0 -> goto C20      /* go 'left' */
       :: hist[i1] == 1 -> goto C21      /* go 'right' */
       fi
    :: (n1 >= n2) && (n1 - n2) < M ->    /* P1 is ahead */
       if                                /* Random choice betw 0 and 1 */
       :: hist[i1] = 0; goto C20
       :: hist[i1] = 1; goto C21
       fi
    fi;
    C20:
       n1 = n1 + 1; i1 = n1 % M;
       vw?[DC] -> vw?DC;
       goto END;
    C21:
       n1 = n1 + 1; i1 = n1 % M;
       xy!RC; goto C1}
  END: skip}

proctype P2()
{ C1:
  atomic{
    tu?[DA] -> tu?DA};
  atomic{
    if
    :: (n2 < n1) && (n1 - n2) <= M ->
       if
       :: hist[i2] == 0 -> goto C20
       :: hist[i2] == 1 -> goto C21
       fi
    :: (n2 >= n1) && (n2 - n1) < M ->
       if
       :: hist[i2] = 0; goto C20
       :: hist[i2] = 1; goto C21
       fi
    fi;
    C20:
       n2 = n2 + 1; i2 = n2 % M;
       vw!DC; goto END;
    C21:
       n2 = n2 + 1; i2 = n2 % M;
       xy?[RC] -> xy?RC; goto C1}
  END: skip}
init { n1 = 0; n2 = 0; i1 = n1 % M; i2 = n2 % M;
       atomic { run P1(); run P2() } }
```

FIGURE 16. Implementing non-local choice using a history variable with bounded length and channels with bounded capacity setting `i`k and `n`k and may only set `hist[`m`]` if `n_`m $\geq$ `n_`j for $k \neq j$ when `i`k= k, otherwise it must follow the decision indicated by the value of `hist[`k`]`.

Figure 16 shows the suggested implementation of the non-local choice example. `P1` sends `DA` as an atomic event. In the next atomic step, `P1` checks whether it is allowed to set the history variable, or whether it has to follow the path determined by `P2` as recorded in the history variable. If `P1` may determine which branch to take in the n-th cycle, it will make a nondeterministic decision.

Implications of the History Variable length. We saw that the capacity of the history variable determines the amount by which processes `P1` and `P2` can 'diverge'. The PROMELA code will only correctly simulate the MSC specification if at all times the difference between the number of the cycle that `P1` is on and the number of the cycle that `P2` is on is $\leq M$. If one process runs ahead of the other by more than M cycles, the PROMELA code will not correctly simulate the MSC specification. A bound on the 'cycle difference' may currently not be specified in MSCs. In fact, as we have noted in previous work, liveness properties such as requiring that a sent message is eventually received are underdetermined by the current standard. It is easy to see that a 'cycle-difference' bound ensures progress of both processes and therefore that this requirement entails a liveness property.

Experimental results. The experimental simulation with XSPIN shows that this implementation of the history variable algorithm satisfies the *n-th-cycle- same-choice* condition but does not prevent the system from blocking. This is in accordance with our semantics and has the following explanation. Consider the following scenario: the system in Figure 14 starts executing, `P1` makes n consecutive `right` decisions, and `P2` is in its m-th cycle ($n > m+1$). Now, `P2` queries the global history variable and follows the `right` decision that `P1` made in the m'th cycle and receives `RC`. In the $m+1$'st cycle, `P2` will again perform a `right` decision, as determined by the global history variable, but find no message `RC` to be received. This is because, as we argued in *op. cit.*, communication in MSCs in non-buffered and therefore $n > 1$ repeated sendings of a message by a given 'MSC arrow' (= message instance) can be received by one receive event. (We retain in the system state a single copy of a message instance that has been sent but not received, and remove this copy when the message is received. A second message-send of an unreceived message instance does not change the system state because the instance is already recorded in the state. One must also be careful to distinguish the *contents* of a message, which may be identified with message type in MSCs, from a message instance. An MSC may contain multiple arrows representing the sending of a single contents. These multiple arrows are different instances of the contents and the MSC state retains a copy of the instance in our semantics.)

6. Introduction of Channels with Capacities

The example in the previous section shows that recording the history of choices that the system makes with respect to non-local choice situations does not suffice to provide a non-blocking interpretation (i.e., a set of execution sequences, none of whom block) of the specification. To provide a non-blocking interpretation, we may add another history variable, a counter variable recording the number of *sendings* and *receivings* of message instances. **[LL95b]** contains simple examples of potential MSC execution sequences during which this variable would be unbounded; this happens when repeated message instances are continually sent faster than they are received, throughout a non-terminating execution. Employing such counting variables while continuing to admit such executions yields an infinite-state system. However, [**LL95a**] argued for the inherent finite-stateness of MSC specifications. So adding such counter variables is not obviously consistent with other requirements on the interpretation of MSCs.

Making Channels explicit. A chief argument against the use of message queues for the interpretation of the communication mechanism in MSCs follows from the *what-you-see-is-what-you-get* (WYSIWYG) requirement on specifications and specification languages. For a visual graphical specification style like MSCs, this means either not adding information that is not explicit in the diagram, or extending the language to capture such information. (We argued in [**LL95b, LL95a**] for an extension to capture underdetermined liveness properties, but for omitting histories since they weren't already explicit. We also noted, and again above, that histories were required in any case reasonably to interpret some already-standardised MSC syntactic constructs.)

This WYSIWYG requirement on specification languages entails that when introducing history variables for messages in MSCs their existence should be made explicit in the graphical representation of MSCs, which means an extension to the syntax defined in the Z.120 standard. Figure 15 shows a possible syntactic representation of channels. We map each message arrow onto one channel. In some situations it may be useful to require different messages to be sent across one channel. This may syntactically be done by attributing channel symbols with name labels, and to understand messages crossing channel symbols with the same name label to be passed along the same channel.

Some Suggested Criteria for Introducing Channels.

1. Channels serve messages following a first-in-first-out strategy.
2. Channels should be free of loss: this requirement may be weakened later.
3. Channels have a specific (finite or infinite) capacity.
4. The semantics of the MSC send primitive should be changed from non-blocking to blocking. Z.120 MSCs have no notion of channels and capacities, therefore there was no point to defining the send primitive as blocking. However, with the introduction of channels and capacities this now makes more sense. Channels with infinite capacity never block on a send, but the blocking becomes important as we introduce channels with finite capacity.
5. The definitions in [**LL95b**] would no longer be appropriate for MSCs-with-channels. However, a semantics can easily be given, in particular by translating an MSC specification into a collection of *Communicating Finite State Machines* [**BZ83**]. Furthermore, a formal operational semantics for PROMELA is currently under development [**NH96**], hence MSCs-with-channels could inherit a formal semantics from the PROMELA translation once this PROMELA semantics is given.
6. The MSCs obtained by introducing unbounded channels cannot be implemented using PROMELA, whose expressive capabilities are limited to finite-state systems.

Benefits. The suggested introduction of channels with finite capacities does not guarantee deadlock-freedom for MSCs[6]. However, we conjecture that blocking due to non-local choices can be avoided by this mechanism. We do not have space

[6]For necessary and sufficient criteria for MSCs to be deadlock-free see [**LS**].

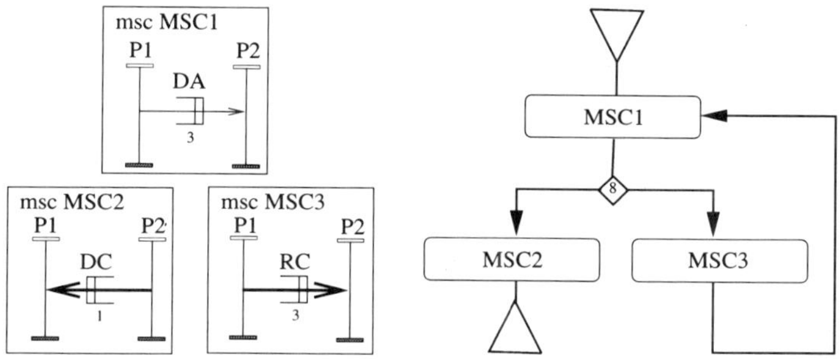

FIGURE 17. MSC example with non-local choice, explicitly represented channels, and explicit capacities for the non-local choice history variable (=8) and channels (=3, 3 and 1)

here for a more-detailed study of how one might introduce channels, and what properties those channels should have. We wished mainly to note how introducing them would solve a problem with 'anomalous' blocking execution sequences. The problem could also be solved by simply accepting the blocking execution sequences as valid executions of the MSC specification.

7. Finite State Implementation

Two of the constructs we have suggested lead to an infinite-state model: non-local-choice *history* variable(s), and *channels* associated with message arrows. However, finite-state-space validation techniques as well as an implementation using PROMELA require a finite-state-space model. Any PROMELA implementation must therefore limit both the capacities of the channels and the length of the history variables for all non-local choice situtations to finite values. We represent these restrictions using appropriate labels on channel symbols and the final-condition symbol leading into a non-local choice situation, as seen in Figure 17. Figure 16 shows the corresponding PROMELA source code.

Making the History Variable Length explicit. Following the WYSIWYG requirement discussed earlier, not only the existence of channels and their capacities should be made explicit in MSC specifications, but also the existence of a history variable and its capacity. Figure 17 shows a possible representation for the presence of a history variable (the diamond shaped connector) and the limitation of its length to 8 (the number in the diamond's interior).

Experiment and limitations. Limiting the capacity of history variables and channel capacities leads to a number of limitations when executing the PROMELA implementation.

1. Figure 18 shows an execution trace generated by XSPIN based on the implementation of the MSC specification in Figure 17 as given in Figure 16. Events in a trace generated by XSPIN are totally ordered (by virtue of their absolute distance to the top of the beginning of each process axis). The left axis corresponds to the PROMELA process generated as a father to processes

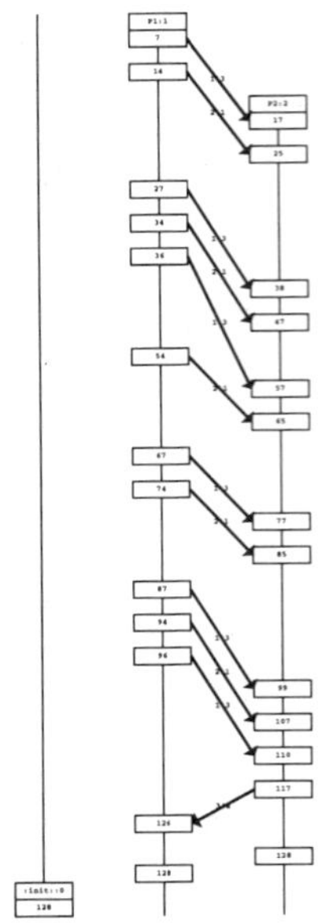

FIGURE 18. Trace of MSC specification generated by random simulation using XSPIN

P1 (middle axis) and P2 (right axis). Note that the maximum divergence between the P1 and P2 is 3 as a consequence of the capacity of the now-blocking communication channel xy, which is 3.

2. As the send primitive is now blocking, processes will not send to a full channel. This excludes a number of interleavings as possible traces of the system. Note that in the implementation without bounded channel capacities a subtrace $(\texttt{!DA, !RC})^4$ could be part of an admissible execution sequence, this is not the case for the example in Figure 17.
3. As argued eralier, the size of the history variable limits the divergence of processes P1 and P2. In the example in Figures 16 P1 could be at most 8 cycles ahead of P2. However, the limitation of the capacity of channel xy is more constraining, limiting the maximal divergence to 3. Both channel capacity as well as history variable length determine the potential divergence of the processes.

8. Verifying Temporal Properties.

The translation of MSCs into a language like PROMELA can prove helpful in two main ways. First, as mentioned above, the translated PROMELA model allows simulation and animation of an MSC specification, which may help in debugging early system designs. Second, as we argued in [**LL95a, LL95b**] MSCs underspecify liveness properties, and temporal logic can be used to specify the additional liveness properties. Desired liveness properties specifiable in linear-time temporal logic

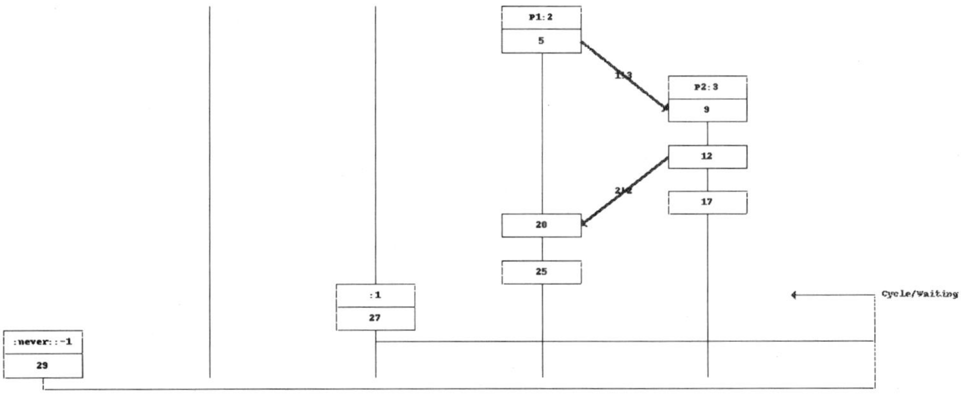

FIGURE 19. MSC showing trace leading to a state violating `[] <> aftsDR`.

(LTL) may be checked in PROMELA, thus PROMELA can support this extension of the MSC language.

It is possible using LTL to specify properties on the PROMELA model generated from an MSC specification that this model cannot guarantee. For example, consider MSC 1 in Figure 12. Suppose we wish to assert that process `P2` will always eventually send a message of type `DR`. XSPIN is a state-based, and not an event-based verifier, so we need to define state predicates specifying the control-state of the process with respect to the events defined in the MSC. Define a state predicate *ta_* (for '*taken*') such that *ta_x* holds iff the last state transition was a sending of a message of type x. The desired assertion is expressed by the LTL formula $\Box\Diamond ta\_\mathtt{DR}$. (We also define and use references to the control state of individual processes in the PROMELA code as XSPIN-LTL propositions.) The property that a message of type `DR` will always eventually be sent cannot hold, because `P2` may eventually decide to execute the left path that describes transmission of a `CC` message followed by termination, along which path the system cannot ever again reach a state in which *ta_*`DR` holds.

The XSPIN environment allows model checking of LTL formulas based on PROMELA models [**GPVW95, Hol**]. LTL formulas can be entered in XSPIN, and a preprocessor translates them into so-called *never* claims [**Hol91, Hol95**]. How can a basic proposition like *ta_x* be defined as a basic proposition in PROMELA? We make use of a predefined PROMELA predicate of the format `process_name[pid]@label_name`. `process_name` is the name of a process as defined in the `proctype` clause; `pid` is a process identification number, generated by incrementing a counter starting at 1 each time a new process of any type is incarnated; `label_name` is a statement label in the PROMELA code. In order to refer to such a predicate, it is assigned a name `pred_name` by a `#define` clause.

In [**LL95c**], the communication event as well as the control-flow branching was included inside a PROMELA `atomic` clause. In order to implement the *ta_* predicate properly using PROMELA labels, we removed the the control-flow branch statement from within the `atomic` clause and labeled it. Figure 21 shows the PROMELA code

```
mtype = {CR, CC, DR};
chan tu = [1] of { byte };
chan vw = [1] of { byte };
chan xy = [1] of { byte };

proctype P1()
{ C1:
  atomic {
    if
    :: tu!CR
    :: full(tu) -> skip
    fi; goto C2};
  C2:
  do
  :: atomic {
       vw?[CC] -> vw?CC;
       goto END}
  :: atomic {
       xy?[DR] -> xy?DR; goto C1}
  od;
  END: skip
}
proctype P2()
{ C1:
  atomic {
     tu?[CR] -> tu?CR; goto C2};
  C2:
  do
  :: atomic {
       if
       :: vw!CC
       :: full(vw) -> skip;
       fi; goto END }
  :: atomic {
       if
       :: xy!DR
       :: full(vw) -> skip
       fi; goto C1 }
  od;
 END: skip
}

init { atomic { run P1(); run P2() } }
```

FIGURE 20. PROMELA code for example 7

implementing the MSC in Figure 12. Labels **aftersCR**, **aftersCC** and **aftersDR** denote points in the process control flow as needed to define *ta_*[7].

As expected, the XSPIN verifier detects a violation of the LTL claim **[] <> aftsDR** which is the translation of the LTL formula $\Box\Diamond ta\_\mathrm{DR}$. For debugging, X-SPIN can run a guided simulation into the state that violated the claim, and the violating trace is illustrated using an MSC (see Figure 19). The online use of this MSC enhances debugging because placing the mouse on individual MSC events highlights the PROMELA code (in another window) corresponding to those events; and it attempts to indicate how the temporal property is violated.

We ran a few more temporal properties through the XSPIN verifier to experiment. Table 1 lists the results. In particular, Property 3 expresses an important

[7]XSPIN automatically generates a 'never'-claim from an LTL and allows for adding the claim to the specification (see [**LL96**] for the code of the 'never'-claim of the LTL formula number 3 in Table 1)

```
mtype = {CR, CC, DR};

#define aftsCR P1[1]@aftersCR
#define aftsCC P2[2]@aftersCC
#define aftsDR P2[2]@aftersDR

chan tu = [1] of { byte };
chan vw = [1] of { byte };
chan xy = [1] of { byte };

proctype P1()
{ C1:
  atomic {
    if
    :: tu!CR
    :: full(tu) -> skip
    fi; }
    aftersCR: goto C2;
  C2:
  do
  :: atomic {
       vw?[CC] -> vw?CC;
       goto END}
  :: atomic {
       xy?[DR] -> xy?DR;
       goto C1}
  od;
  END: skip}
proctype P2()
{ C1:
  atomic {
     tu?[CR] -> tu?CR;
     goto C2};
  C2:
  do
  :: atomic {
       if
       :: vw!CC
       :: full(vw) -> skip
       fi; }
       aftersCC: goto END
  :: atomic {
       if
       :: xy!DR
       :: full(vw) -> skip
       fi; }
       aftersDR: goto C1
  od;
 END: skip}
init { atomic { run P1(); run P2() } }
```

FIGURE 21. PROMELA code including state labels

consistency condition for the protocol represented by the original MSC. This property basically states that once a CC (connect confirmation) has been sent it is not possible to send a DR (disconnect request) afterwards.

9. Summary and Outlook

We noted that simulating MSC specifications had advantages for system designers. We have considered the simulation of MSC specifications in PROMELA, and noted how the questions on the MSC semantics considered in [**LL95b, LL95a**] are reflected directly in the PROMELA executions. We considered in particular verifying properties expressed in LTL. Some liveness properties of MSCs are underdetermined by the standard but any simulation must either allow or avoid each questionable execution sequence, therefore these decisions had to be made. Questions about

Property	Outcome
1. `[] <> aftsCC`	not satisfied
2. `<> aftsCC`	not satisfied
3. `[] (aftsCC -> ! <> aftsDR)`	satisfied
4. `[] (aftsCR -> <> aftsCC)`	not satisfied
5. `aftsCR -> <> (aftsCC \/ aftsDR)`	satisfied
6. `([]<>aftsDR) -> !<> aftsCC`	satisfied

TABLE 1. Temporal Properties verified using XSPIN

non-local choice in MSC branching, which originally arose with conditions but is also present for HMSCs were also considered, and implemented following the development in [**LL95a**]. To avoid imposing over-stringent liveness conditions (that the executing processes may only lag each other by a bounded amount), we considered introducing message-instance-channels and discussed their implementation in PROMELA and some consequences.

Current work includes a formalisation of the MSC-to-PROMELA translation (see [**LL95c**] for a preliminary version) and the development of a tool supporting this translation. Furthermore, we investigate the syntactic analysis of MSC specifications. We have defined syntactic conditions for the occurrence of non-local choices, process divergence and deadlocks in MSC specifications [**BAL96b**]. Combined with the analysis of MSC specifications as discussed in this paper the resulting tool can provide software engineers with substantial support in providing unambiguous first design specifications [**BAL96a**].

References

[AB] Telelogic AB, *SDT*, Participant's Proceedings of the 8th International Conference on Formal Description Techniques FORTE'95, List of tools for demonstrations (G. von Bochmann, R. Dssouli, and O. Rafiq, eds.), p. 455.

[AHP96] R. Alur, G. J. Holzmann, and D. Peled, *An analyzer for message sequence charts*, Tools and Algorithms for the Construction and Analysis of Systems, Lecture Notes in Computer Science, Vol. 1055 (T. Margaria and B. Steffen, eds.), Springer Verlag, 1996, pp. 35–48.

[BAL96a] H. Ben-Abdallah and S. Leue, *Architecture of a requirements and design tool based on message sequence charts*, Tech. Report 96-13, Department of Electrical & Computer Engineering, University of Waterloo, October 1996, 18 p., submitted for publication.

[BAL96b] H. Ben-Abdallah and S. Leue, *Syntactic analysis of message sequence chart specifications*, Tech. Report 96-12, Department of Electrical & Computer Engineering, University of Waterloo, October 1996, 32 p., submitted for publication.

[BMR+96] F. Buschmann, R. Meunier, H. Rohnert, P. Sommerlad, and M. Stal, *Pattern-oriented software architecture - a system of patterns*, Wiley and Sons Ltd., 1996.

[BZ83] D. Brand and P. Zafiropulo, *On communicating finite-state machines*, Journal of the ACM **30** (1983), no. 2, 323–342.

[GPVW95] R. Gerth, D. Peled, M. Y. Vardi, and P. Wolper, *Simple on-the-fly automatic verification of linear temporal logic*, Protocol Specification, Testing and Verification XV (P. Dembiński and M Šredniawa, eds.), Chapmann & Hall, 1995, pp. 1–18.

[Hol] G. J. Holzmann, *What's new in* SPIN *version 2.0*, http://netlib.att.com/netlib/spin/index.html, Version April 17, 1996.

[Hol91] G. J. Holzmann, *Design and validation of computer protocols*, Prentice-Hall International, 1991.

[Hol95] G. J. Holzmann, *The verification of concurrent systems*, AT&T Bell Laboratories, to be published by Prentice-Hall, 1995.

[Hol96] G. J. Holzmann, *Early fault detection tools*, Tools and Algorithms for the Construction and Analysis of Systems, Lecture Notes in Computer Science, Vol. 1055 (T. Margaria and B. Steffen, eds.), Springer Verlag, 1996, pp. 1–13.

[IT95] ITU-T, *Recommendation Z.120, Annex B: Algebraic Semantics of Message Sequence Charts*, ITU - Telecommunication Standardization Sector, Geneva, Switzerland, 1995.

[IT96] ITU-T, *Recommendation Z.120*, ITU - Telecommunication Standardization Sector, Geneva, Switzerland, May 1996, Review Draft Version.

[Jea92] I. Jacobson and et al., *Object-oriented software engineering - a use-case driven approach*, Addison-Wesley, 1992.

[Lab95] NTT Software Laboratories, *SDE technical tour*, Dec. 1995, Presentation slides.

[Leu94] S. Leue, *Methods and semantics for telecommunications systems engineering*, Doctoral dissertation, University of Berne, Switzerland, December 1994.

[LL95a] P. B. Ladkin and S. Leue, *Four issues concerning the semantics of Message Flow Graphs*, , Proceedings of the Seventh International Conference on Formal Description Techniques, Chapman & Hall, 1995.

[LL95b] P. B. Ladkin and S. Leue, *Interpreting Message Flow Graphs*, Formal Aspects of Computing **7** (1995), no. 5, 473–509.

[LL95c] S. Leue and P. B. Ladkin, *Implementing message sequence charts in* PROMELA, Proceedings of the First SPIN Workshop (J.-Ch. Grégoire, ed.), INRS Télécommunications, Montréal, Canada, 1995.

[LL96] S. Leue and P. B. Ladkin, *Implementing and verifying scenario-based specifications using* PROMELA/XSPIN, Participants Proceedings of the 2nd International Workshop on the SPIN Verification System, DIMACS/Bell Labs/INRS-Télécommunications, 1996, pp. 129–146.

[LS] P. B. Ladkin and B. B. Simons, *Static analysis of communicating processes*, To appear, Springer Lecture Notes in Computer Science.

[McM93] K. L. McMillan, *Symbolic model checking: An approach to the state-explosion problem*, Kluwer Academic Publishers, 1993.

[NH96] V. Natarajan and G. J. Holzmann, *Outline for an operational-semantics definition of Promela*, Participants Proceedings of the 2nd International Workshop on the SPIN Verification System, DIMACS/Bell Labs/INRS-Télécommunications, 1996, pp. 175–191.

[OFMP+94] A. Olsen, O. Færgemand, B. Møller-Pedersen, R. Reed, and J.R.W. Smith, *Systems engineering using SDL-92*, Elsevier Science B.V. (North-Holland), 1994.

[RBP+91] J. Rumbaugh, M. Blaha, W. Premerlani, F. Eddy, and W. Lorensen, *Object-oriented modeling and design*, Prentice Hall International, 1991.

[Sel96] B. Selic, *Automatic generation of test drivers from MSC specs*, Tech. Report TR 960514 - Rev. 01, ObjecTime Limited, Kanata, Ontario, Canada, 1996.

[SGW94] B. Selic, G. Gullekson, and P.T. Ward, *Real-time object-oriented modelling*, John Wiley & Sons, Inc., 1994.

DEPARTMENT OF ELECTRICAL AND COMPUTER ENGINEERING, UNIVERSITY OF WATERLOO, WATERLOO, ONTARIO N2L 3G1, CANADA
E-mail address: `sleue@swen.uwaterloo.ca`

TECHNISCHE FAKULTÄT, UNIVERSITÄT BIELEFELD, D-33501 BIELEFELD, GERMANY
E-mail address: `ladkin@techfak.uni-bielefeld.de`

DIMACS Series in Discrete Mathematics
and Theoretical Computer Science
Volume **32**, 1997

Creating Implementations from PROMELA Models

Siegfried Löffler and Ahmed Serhrouchni

ABSTRACT. SPIN is a tool to simulate and validate Protocols. PROMELA, its source language, is a formal description technique like SDL and Estelle that is based on communicating state machines. Unlike most other tools, SPIN is in the public domain and therefore is one of the most widely used formal verification tools today. PROMELA allows to specify distributed automata which can communicate using either message channels or shared memory. This contribution consists of an extension to SPIN which allows the creation of implementations from PROMELA specifications. This can be used for the creation of test scenarios and the rapid prototyping of validated protocol implementations.

1. INTRODUCTION

Formal Description Techniques (FDTs) are used in many fields of software engineering. One main application field is the validation of software designs, especially telecommunication protocols, but they are equally applicable to other fields such as avionics, nuclear power control, medicine, railway control etc. [**C91**] [**J91**] [**HS95**] [**GMPV95**].

Usually the FDTs are used for the validation of a design concept. This validation can be achieved either by simulations of the design or with special validation algorithms. After the concept has shown to fulfill the requirements, the formal specification usually is used as a guideline for the programmer for the creation of an implementation. However it can also be imagined that the formal specification is *automatically* translated into an implementation. This has the advantage of reducing the probability to introduce new faults when translating it.

2. PROMELA/SPIN

The idea to automatically generate executable code from a formal specification is not new. Many efforts have been dedicated to this task in the past [**LCV92**]. The advantage of the PROMELA / SPIN [**H91**] system as basis for an automated generation of implementations is the high data consistency when using SPIN. There is only one tool which is used for the simulation and the generation of the validator.

1991 *Mathematics Subject Classification.* Primary 54C40, 14E20; Secondary 46E25, 20C20.

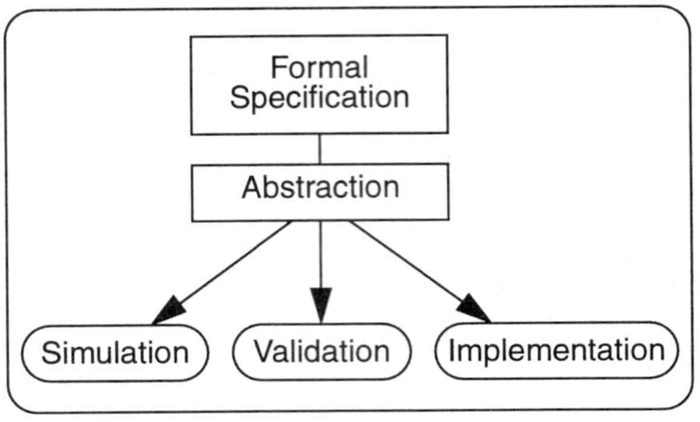

FIGURE 1. Standard Abstract Tree For FDT Tools

By using the same tool, it is less likely that there are inconsistencies between the simulation and the validation. (It is still possible, since SPINs code for the simulation is quite different from the code which is generated as C source code for the validator. In most other systems, independent tools are used for simulation, validation and creation of the implementation. Those tools take the formal specification as a starting point and use an abstract tree like the one shown in figure 1, i.e. a different abstraction is used for simulation, validation and final implementation.

For the simulation and validation, this aspect is not too important. However, for the generation of an an implementation from PROMELA source code the goal obviously is to produce a *validated* implementation which has certain provable characteristics. To achieve this it is necessary to keep as close as possible to the validated code. This means that it is necessary to stay as close to the validated abstraction in the tree in figure 1.

This was our starting point for the generation of the implementation. The validator which SPIN can generate uses a translation of the state machine specified in the PROMELA source code. All transitions in the PROMELA model are translated by SPIN into some lines of C code for the validator. An actual implementation of the model can therefore be realized by reusing this C code representation of the state machine instead of using the PROMELA source code as a basis. In fact Holzmann already mentioned this possibility to create implementations by proposing modifications to the validator as an exercise in his book [**H91**] (p. 317, ex. 13–5). In order to automate this process, it is sufficiant to modify SPIN so that it produces not the original validator but only its state machine part and afterwards automatically generates additional code that is used to execute this state machine.

By modifying the tool SPIN, isolating the code for the generation of the state machine — the "motor" of the validator — and extending it with additional new code (mainly a scheduler), we achieve an implementation which has a very high fidelity to the validated code. The two branches of the abstract tree melt together into one (as shown in figure 2) since the same abstraction is used for validation and implementation.

This is the originality of our contribution that distinguishes it from most other tools for the generation of implementations, which are usually implemented as separate, stand–alone tools (see [**MP94**] for a stand–alone PROMELA to C compiler).

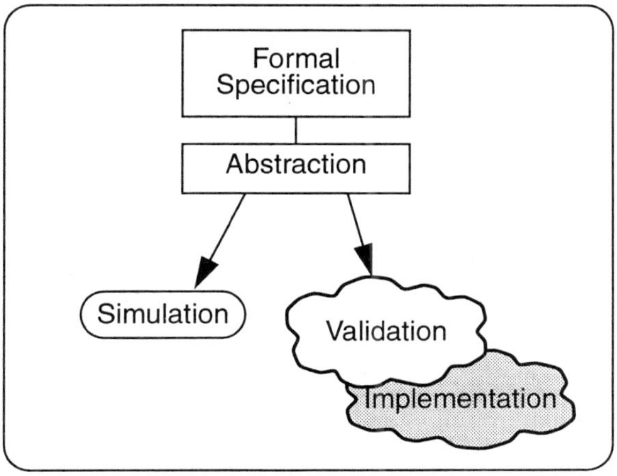

FIGURE 2. Using The Same Tool For Validation And Implementation

3. OUR PROMELA TO C COMPILER

As we already mentioned, the validator which SPIN produces when called with the option "`-a`" is generated as C code. This code is structured in five files, as depicted in figure 3.

In order to re–use the state machine which SPIN generates corresponding to the PROMELA source we need the files "`pan.m`" (the forward moves) and "`pan.t`" (the transition matrix). In addition to those we use parts from "`pan.c`" (the main validator code) and "`pan.h`".

Figure 4 shows an overview over the data dependencies with our extended SPIN tool. The shaded regions in the figure represent the extensions we made for the generation of implementations. This is mainly the scheduler but also some code that allows us to define real–time timers and to communicate with other UNIX processes.

Scheduling in the Implementation

The state machine plus the additional C code for the implementation are compiled into a UNIX program. Thus multiple proctypes which are designed to run in parallel are executed in one single UNIX process. For the switching between the proctypes a scheduler which selects the next active proctype and takes care of external communications and of timers is needed.

Non–Determinism. The most interesting feature of PROMELA as a language is the possibility to describe non–determinism. If non–deterministic choices are to be translated into executables (implementations), there are several possibilities for dealing with it.

In "`if..fi`" and "`do..od`" statements, all *executable* branches from the current state of the state machine can be chosen in an arbitrary manner. Before choosing

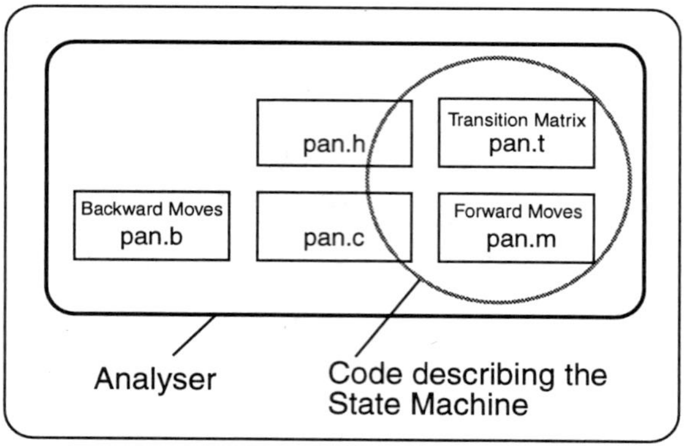

FIGURE 3. Overview Of The Code That SPIN Produces When Called With The Option "`-a`"

a branch, the scheduler therefore has to test if it is executable. The following three strategies for the choice of the next transition can be imagined:

1. It must be allowed to simplify the implementation by just choosing *always the first executable* branch. The scheduler starts with the first transition, checks if it is executable, if it isn't, it tests the second one, and so on. This makes it very compact and usually quite performant. The behavior of the implementation will be the same each time it is started because there is no random element in scheduling (except possible external events).
2. Another possibility is to use a *random number generator* to choose between the branches. The disadvantage of this method is that the same transition branch may be chosen more than once. The scheduler has to keep track of all branches that it already tried to execute because an "`else`" statement that might be one of the branches is only executable if there are no other possible transitions. Since "`else`" is only executable if all other statements in the proctypes state aren't, the scheduler has to do the random branch selections until it has reached all other branches. If a completely random choice is used, some unexecutable branches are probably chosen more than once. Therefore, the scheduler uses more CPU time than necessary.
3. A third possibility is to choose the *first* branch to be executed in a random manner. Afterwards the remaining transitions can be tested sequentially. This costs almost no additional CPU time and has the advantage of introducing a random element into the implementation.

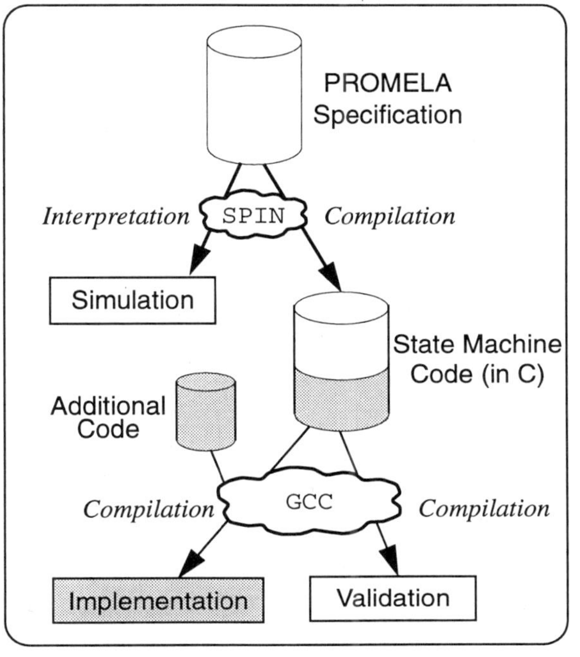

FIGURE 4. Code Dependencies With SPIN

Since we wanted to have at least some random in our implementations, we decided to implement the third algorithm. This also allows us to use implementations for random simulations of the model.

Blocking Proctypes. If there are no possibilities to execute any of the branches in a certain state of a proctype, this proctype should block. If this is true for all proctypes in a UNIX process (or if the current sequence of transitions is "`atomic`") then it is not necessary to check continuously whether a transition has become executable. The only types of events that could change this state of the UNIX process are external messages from other processes or timeouts. So it is possible to block the UNIX process until an external message event occurs or a timer expires. This reduces the CPU load of the machine on which the implementation runs.

Atomic Sequences. In PROMELA, sequences of statements may be defined as "`atomic`" sequences. An atomic sequence should, from the point of view of the other proctypes, be seen as one single instruction that is not interruptible. During the execution of an atomic sequence, the scheduler must not switch to another proctype. However, starting with Version 2.0 of SPIN it is legitimate for an atomic sequence to block [**H95a**] [**H95b**]. In this case it should — since Version 2.0 — be allowed to switch to another proctype.

This is implemented differently in our scheduler. If an atomic sequence in the implementation blocks, the scheduler will stay in this proctype until it unblocks, even if this will never happen.

The reason why we decided to interpret the semantics of the "`atomic`" statement differently is that we wanted to add external events to implementations. However, if an implementation blocks in a certain state because it would have to wait for an external event, one can never know if this event will ever occur or not, because it is not specified in the same model. Therefore we can not know if it is necessary to leave the atomic sequence in order to avoid a deadlock.

A solution to this problem could be to prohibit the use of external channels inside atomic sequences. An easier solution is, in our opinion, to interpret the semantics as they were interpreted by older versions of SPIN, i.e. not allowing the change of proctypes inside atomic sequences at all. If a designer wants to allow a proctype change in a certain state, he should specify this explicitly by breaking the atomic sequence into multiple sequences which are separated by the operation in which the proctype change is to be allowed.

This example shows that although we are using the same abstraction, treat it with the same tool and even use the same internal data structures as in validation, it is still possible to interpret the semantics of PROMELA differently.

Priorities. In Version 2.5 of SPIN, Holzmann added a mechanism to define process priorities for use during random simulations [**H95b**]. Those priorities could be easily implemented in the scheduler, however they aren't yet. The priorities were implemented to improve the debugging facilities. For validation they are not used at all.

Timer Mechanisms

PROMELA was designed as a validation language. For the implementation of a protocol, there are some important requirements that are not yet covered by the language. For example it is absolutely necessary that one can define a timeout. Since the `timeout` statement in PROMELA has initially been designed to avoid the blocking of a proctype, in PROMELA it is not possible to specify any value for a timeout. For the validation it is sufficient to know that a timeout can occur in a certain state. If it can occur, it does not matter how long it may take until this happens since time is an element which does not exist in the validation.

The original `timeout` statement therefore is not supported for the implementation.

A solution to this problem could be to introduce a parameter that can be given to the `timeout` statement. This would have the advantage that existing PROMELA models could be modified very easily. On the other hand, it would have the disadvantage that a change to the language syntax would be necessary.

Since we did not want to change the syntax of PROMELA, we searched for another possibility. As a replacement, a timeout mechanism which uses message queues was implemented. This mechanism is close to the implementation of timers in SDL where timers basically are modeled as messages.

For the communication with this timeout mechanism that is implemented in the runtime scheduler, the following three message channel names, whose names are reserved, were defined:

1. "`set_timer`" This channel can be used to set a timer. The channel definition "`chan set_timer=[1] of { byte, byte };`" has to be used to define the channel in the model. Afterwards, a message consisting of a timer identifier

and a value (in seconds) for the timer can be transmitted in a message to the implementation scheduler.

2. "`timer`" This queue is used by the implementation scheduler to send a message if the timer expires. The channel definition "`chan timer = [1] of { byte };`" has to be added to the PROMELA model. The message consists of the timer identifier that was sent through the `set_timer` channel.
3. "`del_timer`" This queue can be used to delete a timer before it expires. The definition for the channel is "`chan del_timer = [1] of { byte };`" The message should be the timer identifier that was used when setting the timer. In recursive proctypes, it is advisable to use the "`_pid`" Variable to compute a unique timer identifier.

To validate a model which uses those timer queues, an additional PROMELA proctype has to be added to the specification which describes the timer mechanism in the scheduler, i.e. reads and writes the timer queues.

External Communications

We already mentioned that we added external events to the PROMELA models. For the design of the implementations, this results in the necessity to find a way to specify such events in PROMELA. Since we did not want to change the PROMELA syntax by introducing new language constructs, we decided to implement such external events using channels. These "external" channels are specified exactly like normal PROMELA channels, the only difference is that we prefix their names with "`ext_`" in the specification. This allows us to use the existing tools for simulation and validation.

In order to simulate or validate a model which communicates with other systems via external channels, it suffices to include proctypes describing those external events into one single file used for simulation and validation.

The external channels also allow us to divide one PROMELA specification into implementations which run in multiple UNIX processes and communicate with each other via external channels. For this we have to create a PROMELA source code for each UNIX process to be generated. Using "`#include`" preprocessor directives all proctypes can then be included either in the source file which is used for simulation or in the source files which are used for the compilation of the implementation.

The connection of the UNIX processes is realized in a client–server architecture. The server always keeps track of the queue contents of the external channels in all connected UNIX processes. If a client wants to read from an external channel, it has to send a request to the server. On the other hand, the server notifies all clients if the contents of any of the external channels changes. To avoid having access problems with two clients reading from the same channel we limited read access to an external channel to one UNIX process per channel. If a UNIX process has read from a channel, thereafter no other UNIX process is granted read access to the same channel. This reduces the communication between the UNIX processes and makes them much faster. Nevertheless, multiple proctypes are allowed to write into the same channel queue.

For the inter–process communication between the UNIX processes, AF_UNIX domain sockets are used. This makes it easy to distribute the processes over multiple machines by changing the AF_UNIX domain into the AF_INET internet domain.

Upon compilation of a PROMELA model, one can specify whether the compiled UNIX process should be the server or a client by setting the appropriate switch. Since all channel queues are kept within the server, it is advisable to make the UNIX process the server which uses the external channels the most.

4. Future Work

The most important drawback in our extended SPIN environment is that it is not yet possible to use synchronous channels for the communication between UNIX processes. Almost as important is the missing capability for "sorted send" and "random receive" operations between UNIX processes, which were added to PROMELA with version 2.0 of SPIN. [**H95a**] [**H95b**]. Also introduced with SPIN Version 2.0 and not yet supported are some other language constructs like the "`d_step`" statement.

It has to be mentioned that the compiler itself is not validated and most probably still contains quite a lot of bugs. Therefore the resulting implementations are not 100% validated.

Another thing which is left to do is a validation of the protocol that we use for the connection of clients and servers. A basic description of this protocol can be found in [**L96**].

5. Conclusions

With PROMELA / SPIN, Holzmann has attacked two of the main factors inhibiting more widespread use of specification or validation tools, namely difficulty of use and the inherent limitations of the finite state reachability methods.

However, for the step to the final implementation, some difficulties remain. The major drawback of PROMELA in its current state is that the semantics are still not exactly specified, therefore allowing interpretations which may lead to problems when implementing the design. Another problem is the missing capability of the language to refine the specification in a way that suffices to describe details (like timeouts) of an implementation.

Our extended SPIN tool is usable for the rapid prototyping of validated implementations of communication protocols. Because of our different interpretation of the "`atomic`" statement, it is possible for the implementation to behave not exactly as expected. The generated implementation is not 100% validated because we had to add code for a scheduler and for external communications which has not been exhaustively verified.

The main application field we see for the implementations generated with the current version of our extended SPIN tool is the rapid prototyping of testing scenarios.

References

[H91] Gerard J. Holzmann, *Design and Validation of Computer Protocols*, Prentice Hall, 1991.

[H95a] Gerard J. Holzmann, *What's New in* SPIN *Version 2.0*, in "SPIN Documentation", 1995 `http://netlib.bell-labs.com/netlib/spin/whatispin.html`

[H95b] Gerard J. Holzmann, *V2. Updates*, in "SPIN Documentation", 1995 `http://netlib.att.com/netlib/att/cs/home/holzmann-spin.html`.

[LCV92] A. A. F. Loureiro, S. T. Chanson and S. T. Vuing, *FDT Tools for Protocol Development*, Forte '92, Lannion, France, 1992.

[NO95] E. Najm, F. Olsen, Télécom Paris, *Reactive* PROMELA, 1st SPIN Workshop, Montreal, 1995.

[MP94] Eric Moreau, Jérôme Paillet, PROMELA *Compilateur — Document Technique*, Rapport de Stage, Télécom Paris, 1994.

[L96] S. Löffler, *A* PROMELA *To C Compiler, Rapport de Stage*, Télécom Paris, 1996 `http://www.res.enst.fr/~loeffler/abstract.html`.

[C91] J. A. Chaves, *Formal Methods At AT&T - An Industrial Usage Report*, Forte 91, Sidney, 1991.

[J91] C. Jones, *Formal Methods And Their Role In Industry*, ASWEC 91, 1991.

[HS95] R. P. Hautbois, P. de Saqui–Sannes, *Results And Viewpoints On The Use Of Formal Languages*, Workshop "Formal Methods, Modelling And Simulation For System Engineering", St–Quentin En Yvelines, France, 1995.

[GMPV95] R. Groz, J. F. Monin, M. Phalippou, D. Vincent, *Current Application Of Formal Methods In France Télécom — CNET*, Workshop "Formal Methods, Modelling And Simulation For System Engineering", St–Quentin En Yvelines, France, 1995.

The first author is also affiliated with the University of Stuttgart, Germany.

SIEGFRIED LÖFFLER, ECOLE NATIONALE SUPERIEURE DES TELECOMMUNICATIONS, DEPARTEMENT RESEAUX, 46 RUE BARRAULT, 75634 PARIS CEDEX 13, FRANCE

E-mail address: `floeff@tunix.mathematik.uni-stuttgart.de`

AHMED SERHROUCHNI, ECOLE NATIONALE SUPERIEURE DES TELECOMMUNICATIONS, DEPARTEMENT RESEAUX, 46 RUE BARRAULT, 75634 PARIS CEDEX 13, FRANCE

E-mail address: `ahmed@res.enst.fr`

DIMACS Series in Discrete Mathematics
and Theoretical Computer Science
Volume **32**, 1997

Modelling and Verification of the MCS Layer with Spin

Pedro Merino and Jose-Maria Troya

Abstract. This paper presents the first results of work aimed at using the Spin tool for the verification of multi-party protocols. We focus on the ITU-T Multipoint Communication Service, a generic service designed to support highly-interactive multimedia conferencing. We have built a formal Promela model of the protocol recommended for this layer, and have described some requirements using linear-time temporal logic. Next, we have verified this model against these properties using the model-checking capabilities of Spin. This analysis demonstrates that most of the properties are satisfied, but also that the interpretation of the standard documents could introduce errors in a real implementation. Apart from the verification of the protocol itself, the aim of the work is the evaluation of Promela and Spin for multi-party protocols such as the ones employed for multiconferencing or collaborative environments. In this case, we detect some deficiencies in Promela for modelling certain aspects of the protocol behavior and also for formalising various temporal requirements. We have suggested some extensions to improve the language.

1. Introduction

The recommended starting point to achieve reliable communication software is a validated and verified formal specification of the protocol to be implemented. Validation involves ensuring that the formal description of the protocol satisfies certain predefined correctness criteria such as absence of deadlock, unspecified receptions or unreachable code. Verification of protocol specific constraints, usually related to sequences of states, involves comparing the description of these requirements against a description of a protocol proposed to satisfy them. Several languages and tools have been designed for the computer aided analysis of both kinds of properties. We are interested in the use of the language Promela and the Spin tool for analysing multi-party communication protocols [**H91, H95**].

Most communication protocols are point-to-point, with only two communicating parties. The emergence of collaborative environments, however, has stimulated the development of multipoint protocols, where several nodes communicate with other simultaneously. The set of nodes, connections and node functions in such systems is dynamic rather than static, and simultaneous activity is the essence of the interaction. This degree of interaction could produce more global states and

---

1991 *Mathematics Subject Classification.* Primary 68Q60; Secondary 68N99, 68U30.
This work was supported in part by Project CICYIT TIC94-0930-C02-01.

events during the execution of the system, and also more unpredictable behaviors than have been considered by the designer. To automatically verify these systems, formal specification languages require additional mechanisms to model the behavior and properties.

PROMELA and SPIN seem to offer suitable mechanisms for expressing the behavior and properties of these kinds of concurrent systems. Protocol behavior is modelled as communicating finite state machines extended with variables, and with the dynamic creation of new machines. Temporal properties can be expressed as Büchi automata over some observable events in the behavioral part [**P77**]. SPIN offers translation from Linear Time Temporal Logic (LTL) [**ITU93b**] formulas into PROMELA Büchi automata, [**GPVW95**] so the user can choose between two ways of representing the requirements. Validation of general properties is performed by reachability analysis [**W78**], while on-the-fly model-checking is employed for the verification of temporal properties [**CE86, GPVW95**]. We believe that the tool could be improved further to deal with multi-party protocols.

In order to evaluate SPIN for the modelling and verification of multi-party protocols, this paper presents the first results of analysing the multipoint communication service (MCS) defined in the ITU-T T.122 and T.125 recommendations [**ITU93a, ITU93b**]. MCS is a multi-party communication service designed to support highly-interactive multimedia conferencing applications, and is based on the concept of a domain established over a collection of point-to-point connections. To create an MCS domain, an MCS provider establishes a connection to a remote provider and binds this connection to a domain. The complexity of the domain can vary from a simple point-to-point connection to a multi-branched tree structure of connections. Application clients request their local providers to attach them to one or more domains in order to exchange data with other clients in the same domain. Due to the tree structure, the attachment of a client to a domain causes intermediate providers to route control messages up and down the tree. This part of the protocol, connection management and client attachment, seems to be a suitable candidate to be analysed with an automatic verifier to ensure that there are no unexpected sequences of events.

Our analysis reveals some potential problems in the interpretation of the T.125 document, and also some deficiencies in the SPIN tool. We suggest how the protocol problems can be solved in a real implementation. In addition we propose certain extensions to PROMELA. The first idea is dynamic channel allocation for allowing a more direct representation of unbound networks of processes, which are dynamically created. Other extensions are syntactic manoeuvres to simplify the definition of propositions for temporal properties.

The organisation of the rest of the paper is as follows. First, in section 2 we briefly review the service and protocol specification of the MCS layer. In section 3, the PROMELA model of the behavioral part and the LTL formulas for properties are presented. In section 4, we explain current results concerning the verification itself. Section 5 details the suggested extensions to PROMELA. Finally, section 6 presents the conclusions and outlines further work.

2. The MCS Layer

2.1. Overview. ITU-T recommendation T.122 defines this service as a multipoint delivery service, providing mechanisms for multiple applications for sending

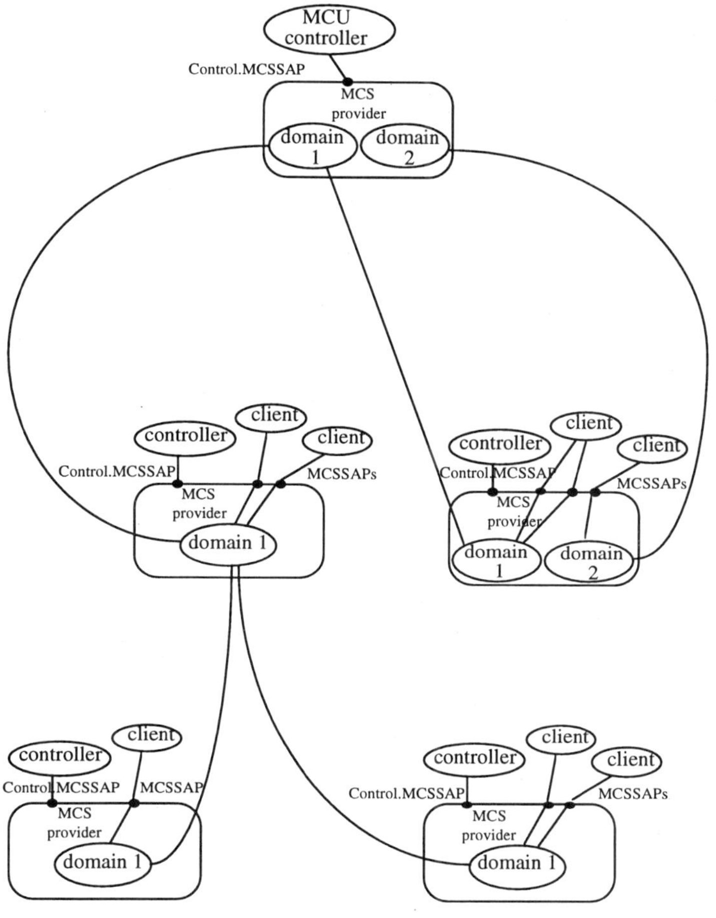

FIGURE 1. Two MCS Domains.

data to every destination or to some destinations within a group using a single primitive. Multipoint communication is supported with multipoint channels, to which client applications are attached. It also provides a token mechanism to allow applications to control limited resources or to carry out synchronisation. Figure 1 represents a typical configuration with several clients using the service. The service is implemented with an MCS provider in every node.

This service is organised around hierarchical domains, grouping a set of clients which send and receive data within the domain. One client can be part of different domains. Domain management implies the use of primitives for creating connections between computers in the domain, attaching and detaching clients in a domain, and splitting domains (see Figure 2). Construction and deconstruction of the domain is the responsibility of a special user process called a controller. A domain is created when two controllers agree on a new connection. Every node which provides the MCS service needs a controller. A given node in the network can be configured as a multipoint control unit (MCU).

2.2. Connection management. A domain is composed of a set of connections between each pair of nodes. An MCS connection consists of one or more transport connections between the same pair of MCS providers. The number depends on the data transfer priorities to be implemented. If only one priority is implemented in an MCS domain, a minimum number of messages for connections are necessary. The MCS layer maintains the integrity of its connections in a domain. To ensure this, the connections are oriented with an up side and a down side. Every domain has a top node, which is responsible for controlling user attachment, the use of channels, and the control of tokens in its hierarchy.

An MCS connection merges two previous domains, and can produce a node to become the top node of the domain. In this case, a maximum height in the domain tree is permitted, in such a way that MCS connections are removed when required. When an MCS connection is released, the domain is split and the portion containing the top provider survives while the other eradicates itself. Merging and splitting domains updates the channels in use.

2.3. User Attachment. Every client receives a unique identifier when it attaches itself to a domain, the identifier being a particular class of channel identifier. The responsibility for maintaining user identifiers in a domain resides in the top domain, so requests have to travel from the clients local node until the top node is reached. Then the reply message travels along the same path in the opposite direction, and every provider in the path records the new user identifier in a local information base.

The attachment of new users is particularly problematic if this is attempted when the domain is being expanded, because merging implies the confirmation of various user identifiers in the domain and the purge of others. A provider finding itself at the lower end of a connection starts the merging process. Since it is unsafe to transact user requests while the information base is being updated, these requests are safeguarded and processed later.

3. The PROMELA Model

3.1. The Behavior. Although the scheme can be further refined to include new functions, in this model we only consider domain creation and destruction (*mcs_connect_provider, mcs_disconnect_provider* primitives) and user attachment and detachment (*mcs_attach_user, mcs_detach_user* primitives.) In order to implement these primitives, we have to handle the following messages, according to T.125: *connect-initial, connect-response, pdin, edrq, mcrq, mccf, pcin, dpum, aurq, aucf, durq and duin.* Figure 2 and 3 summarises the meaning of primitives and messages considered in the model.

Besides simplifying the model to make a more practical analysis, we have considered following assumptions, some of which are motivated by ambiguities in the T.125 document:

- MCS connections are simplified by taking into account only one transport connection, and supporting only one priority in the data transfer.

Primitive	Function
mcs_connect_provider (request, indication, confirm, response)	Create an MCS connection to extend a domain.
mcs_disconnect_provider (request, indication)	Remove an MCS connection in a domain.
mcs_attach_user (request, confirm)	Attach a client to an existing domain
mcs_detach_user (request, indication)	Remove a client from a domain.

Message	Function
connect_initial	Ask the creation of a new MCS connection. It is generated by a mcs_connect_request.
connect_response	Confirm a MCS connection. It is generated by a mcs_connect_response.
pdin	Plumb a hierarchy of MCS providers below a new MCS connection to ensure that no cycle has been created. It is also generated by the top MCS provider to enforce the maximum height of a domain.
edrq	Communicate upward changes in the height of providers. It is generated at the lower end of a connection wherever its height changes.
mcrq	Communicate upward the user ids held by a former top provider so that they will be incorporated in the merged domain.
mccf	Reply a preceding mcrq. User ids not incorporated into the merger domain are reported as channel ids to be purged. Accepted user ids are reflected into the information base of intermediate providers.
pcin	Purge the use of user ids from subordinate providers. It is broadcasted downward from a former top provider following receipt of mccf.
dpum	Compel the receiver to disconnect the MCS connection.
aurq	Request the assignment of a new user id. It rises to the top provider, which returns an aucf reply.
aucf	Confirm the assignment of a user id.
durq	Delete the user id from the information bases. A duin is broadcasted to advise other providers of the change. It is generated by an mcs_detach_user request, and also when a MCS connection is disconnected.
duin	Indicate the elimination of a client in the domain. The client itself ceases to exist when receipt a detach user indication containing its own user id.

FIGURE 2. MCS Primitives and Messages in the model.

- There are no application clients attached to a top provider, until the provider becomes non-top.
- Two bounded queues are used in every intermediate MCS provider to temporarily store the messages *aurq* and *mcrq* that have to be confirmed.

- A *pdin* message is sent by a MCS provider only when it ceases to be a top provider in the domain.

Figure 3 represents the ideal decomposition of the MCS provider into processes. Arcs represent unidirectional channels. Instead of writing monolithic code, the main functions of the MCS provider are implemented with two kinds of processes: control and domain. The responsibility for maintaining each provider's integrity corresponds to the former, while instances of domains are created to manage different domains in the node. The control process deals with primitives for MCS connections and with *connect_initial* and *connect_response* messages. When a connection is successfully created between two control processes, they create the corresponding domain process, which manages the rest of the message as in Figure 3. If the domain process is already running, then the control sends an internal *new_mcs* whenever a new connection is created. The domain informs the control when connections are removed due to protocol execution.

The process *mcs_sap* is designed to prevent a client from sending a message *mcs_attach_request* to a non-existing domain. These primitives are processed by the control process, which is responsible for notifying the control processes about new clients attaching in the node. The multiport processes are employed for maintaining separate channels among processes, while keeping the code independent of the number of processes. Finally, application level processes simulate simplified versions of controllers and clients to initiate the creation of MCS connections and the attachment of user programs.

Our current high-level specification is still simpler, because we are only working with one domain and one client per node. We have considered five nodes to create the domain and communication channels with a capacity of two. The internal queues for *aurq* and *mcrq* messages have been modelled with PROMELA channels with a capacity of five. The current model contains about 500 lines.

3.2. The properties. SPIN enables the expression of properties using PROMELA statements such us assertions, progress labels, acceptance labels and never claims. The most complex method is the never claim combined with acceptance labels, which have the same expressiveness as Büchi automata. The tool also offers an on-line translation of LTL formulas to never claims, in such a way that the designer can express complex properties without worrying about the semantics of the internal automata representation.

The major function of the MCS provider is to maintain the integrity of the MCS connections comprising an MCS domain. An the other basic function of the modelled part is to uniquely identify users attached to a domain. Both global requirements could be expressed with the following properties:

- There is a single MCS provider at the top of each domain
- The height in every domain is always bounded
- The protocol always detects and removes cycles in MCS connections
- Every client attached to a domain has unique identifier

It is clear that if we construct LTL formulas with the form []p, p being the proposition to be satisfied, none of the properties will be satisfied in a real scenario. The reason is that the final structure of the domain is only achieved after several steps of merging previous groups of nodes, each one with a top node and several users. So, we define a safe domain as one satisfying two characteristics:

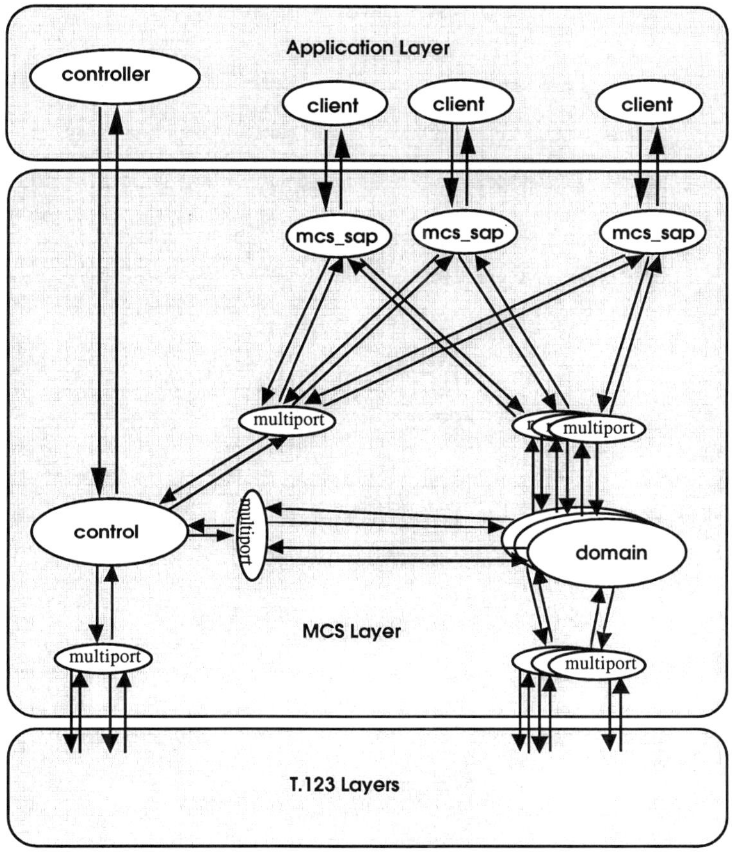

FIGURE 3. Extended model of MCS provider.

- all the nodes to be part of the domain are already connected by MCS connections.
- there is no MCS provider still merging domains.

A merging domain, in contrast, is one that satisfies only the first of these two constraints.

We now describe all the properties in an informal way:

- P1: In a safe domain there is a single MCS provider at the top of each domain.
- P2: In a safe domain the height in every domain is bounded.
- P3: In a safe domain every client attached to a domain has unique identifier.
- P4: In a merging domain, the protocol always detects and removes cycles in MCS connections (it will become safe).

Property	LTL formula	LTL formula to produce *never claim*
P1	[ ] (safe -> one_top)	<> (safe && !(one_top))
P2	[ ] (safe -> bounded_height)	<> (safe && !(bounded_height))
P3	[ ] (safe -> unique_ids)	<> (safe && !(unique_ids))
P4	[ ] (merging -> <> safe	<> [ ] merging
P5	[ ] (merging -> ! (new_id U safe))	<> (merging && new_id)

FIGURE 4. LTL formulas for the properties.

- P5: While a domain is in a merging state, no new user identifiers are assigned.

Due to the semantics of the never claim and the way that SPIN carries out model-checking, the best way to verify these properties is to formalise them as errors, and then try to find an execution sequence of the protocol matching the error. Figure 4 represents properties 1 to 5 using LTL formulas for the (positive) requirements, and an alternative form to express error conditions.

In the case study of five nodes trying to form a single domain, we have constructed the model with a set of global variables which have been employed to define the propositions in the formulas.

4. Results

The MCS model has been analysed by random simulation and by exhaustive validation using Spin Version 2.8.4 . Every LTL formula was translated into a never claim and then verified. The size of the state vector was about five Kbytes. After removing the programming errors detected with the tool, the formal specification is free from general errors, and also satisfies most of the temporal properties. The only property that was problematic is property P2.

The problem with this property is that control of the maximum height is always initiated by the down side in a new MCS connection by sending the *puin* message. The initial value of height in the message is the correct one to prevent cycles, but it never detects a height violation from the older top provider. This control could be initiated in the new top provider which has been attached to the previous top with the new connection. But this point is not clearly specified in the T.125 recommendation.

One suggestion to prevent this problem could be to broadcast a *pdin* message from the top provider if processing the *edrq* produces an illegal value for its local height. This message will eventually produce the elimination of one or several MCS connections in the tree, and new *edrq* messages will go to the top as an effect of disconnection.

5. Extending PROMELA

Modelling the Behavior. PROMELA offers suitable mechanisms to model properties of certain selected functions MCS layer. Improvements could be made to avoid some tedious details when the configuration of the system is dynamic rather than static.

One of the problems in modelling the dynamic creation and termination of processes is the allocation of channels for connecting them. The current version of the language allows only for the definition of new channels as local variables in new processes, but it is not possible to create two processes and explicitly connect them by means of a new channel. This implies that a maximum number of channels has been previously declared, and the programmer is responsible of using them as a pool, allocating and deallocating them when necessary. This method unnecessarily increases the size of the state vector. We suggest solving this problem in the following way:

Dynamic allocation of channels:
c = allocate(size, {type1, type2, ...})
to initialise a channel c, c being a variable declared as chan c;
free(c)
to remove the contents of channel c.

```
Example:
        proctype p1(chan  in)
        {  ............
           /* process messages  until receive end */
           ...........
           free(in)
        }
        proctype p2(chan  out)
        {  ............
           /* put a set of messages into channel out  */
           ..........
        }
        init
        {  int i; chan c;
           do
           :: i == max -> break;
           :: else ->
               c = allocate(2, {mtype, int});
               atomic { run p1(c); run p2(c); }
               i++;
           od
        }
```

Expressing temporal properties. Another problem is the definition of propositions to construct temporal formulas or never claims. Again, the dynamic nature of the system involves an unnecessarily large code. We propose several extensions to be included in never claims.

References to labels:

`p[any]@label` True if any process of type p reaches the label

References to arrays:

`a[any]` *relation* `value` True if a is an array and some position of a satisfies *relation*.

`a[every]` *relation* `value` True if a is an array and every position of a satisfies *relation*.

`a[one]` *relation* `value` True if a is an array with exactly one position satisfying *relation*.

6. Conclusions

This paper presents the current state of our work on the modelling and verification of communication protocols for collaborative environments. As a first step in the evaluation of PROMELA and SPIN for this purpose, we have analysed part of a Multipoint Communication Service layer. A model of this layer was constructed, and several temporal formulas were defined to verify some potential problems in the standard documents.

The analysis demonstrates that the tool is suitable for detecting errors in the specification, in many cases due to ambiguities in the standard. Some refinements could facilitate the use of PROMELA for multi-party protocols, which are characterised by a dynamic configuration. Several language extensions have also being proposed.

Current work focuses on the implementation of the language extensions by preprocessing filters, and on the verification of a more complete model of the protocol.

References

[CE86] E. M. Clarke, E. A. Emerson, A. P. Sistla, Automatic Verification of Finite-State Concurrent Systems Using Temporal Logic Specifications. *ACM Trans. on Programming Languages and Systems*, 8(2), 1986.

[GPVW95] R. Gerth, D. Peled, M. Y. Vardi, P. Wolper, Simple On-the-fly Automatic Verification of Linear Temporal Logic. *Proc. PSTV Conference*, 1995.

[H91] G. J. Holzmann, *Design and Validation of Computer Protocols*, Prentice-Hall, 1991.

[H95] G. J. Holzmann, *What's New in* SPIN *Version 2.0*, Bell Laboratories, 1995.

[ITU93a] ITU, *ITU-T Recommendation T.122. Multipoint Communication Service for Audiographic and Audiovisual Teleconferencing*, ITU (Geneva), 1993.

[ITU93b] ITU, *ITU-T Recommendation T.125. Multipoint Communication Service Protocol Specification*, ITU (Geneva), 1993.

[P77] A. Pnueli, The temporal logic of programs, *Proc. 18th IEEE Symposium on Foundation of Computer Science*, IEEE, 1977.

[VW86] M. Y. Vardi, P. Wolper, An Automata-Theoretic Approach to Automatic Program Verification *Proc. Symp. on Logic in Computer Science*, 1986.

[W78] C. H. West, General Technique for Communication Protocol Validation, *IBM J. Res. Develop*, 22(4), p. 393, 1978.

DEPARTAMENTO DE LENGUAGES Y CIENCIAS DE LA COMPUTACION, UNIVERSIDAD DE MALAGA, 29071 MALAGA, SPAIN

E-mail address: `pedrolcc.uma.es troyalcc.uma.es`

DIMACS Series in Discrete Mathematics
and Theoretical Computer Science
Volume **32**, 1997

Protocol Verification with Reactive PROMELA/RSPIN

Elie NAJM and Frank OLSEN

ABSTRACT. REACTIVE PROMELA/RSPIN is an extension to the protocol validator PROMELA/SPIN. It enhances the simulation and verification capabilities of SPIN by allowing modular specifications to be analysed while alleviating the state-space explosion problem. REACTIVE PROMELA is a simple reactive language. The tool RSPIN is a preprocessor for SPIN which translates a REACTIVE PROMELA specification into a corresponding PROMELA specification. The main function performed by RSPIN is to combine *configurations of* REACTIVE PROMELA *automata* into PROMELA *proctypes.* The translated specification can then be simulated and verified using SPIN.

We demonstrate the language and tool by the specification, translation, simulation and verification of the LAP–B data link protocol. This protocol is quite complex, and benefits from decomposition.

1. Introduction

When considering the problem of specification, (de)composition is a central issue. PROMELA provides for two styles of composition of automata: loosely coupled (communication by FIFO queues) and tightly coupled (communication by rendezvous). A third style, the synchronous reactive style, has been widely advocated and used in the literature and in industry.

In the synchronous reactive style, a configuration of automata reacts to external events in a synchronous way: a collection of external events is treated thoroughly by the configuration before another collection of events is taken and processed. In other words, the reactive configuration reacts to input events, and it is only at the end of the reaction that new inputs can be considered and processed. This kind of processing is valid when the speed of a reaction is higher than the delay between two consecutive input events.

The reactive style allows for powerful decomposition of specifications, beyond what is possible with merely rendezvous between automata. Furthermore, it reduces the state space explosion by constraining parallelism between automata.

Whereas Holzmann in [**Hol91**] proposes ways of reducing the complexity of systems (by incremental composition, minimization, generalization, atomic sequences, layering and structuring techniques, and so on), this is not a feature of the PROMELA

1991 *Mathematics Subject Classification.* Primary 68Q60; Secondary 68N99, 68U30.
Frank Olsen was supported by a grant from CNET.

language in itself. Instead it is a guideline for how to use the language for large, complex systems.

This paper describes an extension to PROMELA, whereby reactive processes can be defined and instantiated. A reactive process is a configuration of synchronously composed automata. Besides the linguistic extension, this paper also describes a translation mechanism of reactive processes into PROMELA processes. This translation has been implemented in a preprocessor to SPIN, called RSPIN which translates a specification in REACTIVE PROMELA into an equivalent one in PROMELA.

We illustrate the language and the tool with the specification and validation of LAP–B protocol. We will show that errors in the specification can be detected by using the SPIN tool to perform exhaustive validation.

1.1. Organisation of the paper. The paper starts with an overview of Gerard Holzmann's PROMELA language and SPIN tool in section 2. It is followed, in section 3, by an informal description of REACTIVE PROMELA, our proposed extension to PROMELA and its associated tool RSPIN. In section 4 we show how we have used REACTIVE PROMELA to specify and verify the LAP–B protocol. We have also included the formal semantics of REACTIVE PROMELA and of a subset of PROMELA in appendix A. For completeness the full REACTIVE PROMELA code for the LAP–B protocol is given in appendix B.

2. Overview of PROMELA/SPIN

In this section we give a brief overview of the PROMELA language and the associated SPIN tool. More information can be found in [**Hol91**].

PROMELA is a language used to model communication protocols, and other kinds of distributed systems, at an abstract level. A program in PROMELA consists of processes that communicate either asynchronously over FIFO channels or by binary rendez-vous between processes. The processes are extended Finite State Machines (EFSMs). Only simple data–types are available: integers of various ranges and record types (similar to typedefs in the C language). The reason is that code–generation is outside the scope of PROMELA: it is a pure *validation language.*

PROMELA does not have a formal semantics, although some authors have given the semantics of a subset of PROMELA: one was in our paper at the TACAS'96 conference [**NO96**]; another, also presented at TACAS'96, was given by S. Tripakis and C. Courcoubetis in [**TC96**].

Associated with the language is a tool called SPIN. It allows PROMELA systems to be simulated, either step-by-step or randomly. Furthermore it is possible to perform an exhaustive simulation by generating the complete state-space of the system. Properties like deadlock and livelock can be detected automatically by checking the generated state-space. Detection of these properties rely on special labels (*end–state-*, *acceptance-*, and *progress* labels) that are placed in the PROMELA code by the user. Even more interesting is that other more specific properties can be expressed in linear time temporal logic, and be verified by exhaustive simulation. Unfortunately only one temporal logic formula can be verified at a time which is not convenient. All properties are verified *on–the–fly*, i.e. it is not necessary to generate the whole state space to detect errors, (although to verify that the specification is error–free it is required to generate the whole state–space).

The verification part of SPIN uses several different state–space generation algorithms, including normal depth–first search and Holzmann's *Supertrace* bitspace algorithm. More recent versions of Spin uses partial–order methods [**God94**] to perform exhaustive simulation on a reduced state-space. The aim of this is to combat the state-explosion problem which limits the size of systems that can be verified.

3. Overview of Reactive Promela/Rspin

We now go on to describe our Reactive Promela, our extension to Promela. It aims at enhancing Promela in the following areas:

- **Decomposition:** Promela gives the user a single level of decomposition: the process level. We add a *reactive process* which can be decomposed into a set of communicating *automata.*[1]
- **Reactive composition:** Promela processes can be composed in parallel via asynchronous message channels or by binary rendez–vous. The automata that make up a reactive process communicate using a *synchronous reactive* style of communication[2]
- **State–space reduction:** the automata that make up a reactive process are combined into a single Promela process using the Rspin tool. If these automata had been modelled as Promela processes the Spin tool would have created the full interleaving of their actions with all other processes on the specification. Our combination algorithm prevents this interleaving by forbidding external inputs during a reaction.
- **Atomic reactions:** when a reactive process accepts an input it reacts atomically by changing its state and by generating a set of outputs. This is simpler than using Promela's atomic sequence statement, since the Rspin tool automatically encapsulates reactions with this construct. Hence, the user does not need to decide in each case whether to use atomic or not.

This section is organised as follows: we first give the concrete syntax for Reactive Promela in section 3.1, before we give an informal overview of the language, expanding on the above introduction, in section 3.2.1–3.2.6. Then in section 3.4 we discuss the accompanying Rspin tool.

3.1. Syntax of Reactive Promela**.** The syntax of Reactive Promela strongly resembles that of Promela, since the aim is to make it as easy as possible to use the extension. The only new keywords added in Reactive Promela are the following:

```
automaton  in         inport    link
outport    rproctype  external
```

Below we present the parts of the Reactive Promela grammar where it extends the Promela grammar. First, a few words on the notation. The new keywords are displayed in capitals (RPROCTYPE), tokens and Promela keywords are enclosed within apostrophes (':' and 'goto'), names (references) are displayed in lowercase letters within < ... > (<rproc_name>) and non-terminals in lowercase letters (r_proc). Also, { ... }+ means one or more of the enclosed unit and

[1]We also consider providing arbitrary levels of decomposition.

[2]Note that neither reactive processes nor Promela processes communicate using the synchronous reactive style, only automata do.

`{ ... }*` means zero or more units. Units enclosed by `[ ... ]` are optional.

In Reactive Promela, the old process definition:

```
proc ::= PROCTYPE ...
```

is replaced by:

```
proc ::= p_proc | r_proc
```

where `p_proc` is the usual Promela process, and `r_proc` is the Reactive Promela process defined by:

```
r_proc      ::= RPROCTYPE <rproc_name>
                '(' r_interface ')' r_body
r_interface ::= {r_port_decl}*
r_port_decl ::= INPORT <port_name> | OUTPORT <port_name>
r_body      ::= {automaton}+ links
automaton   ::= AUTOMATON <autom_name>
                '(' a_interface ')' a_body
a_interface ::= {a_port_decl}*
a_port_decl ::= EXTERNAL INPORT  <port_name>
              | EXTERNAL OUTPORT <port_name>
              | INPORT  <port_name> port_init
              | OUTPORT <port_name> port_init
port_init   ::= '=' '{' type_list '}'
a_body      ::= '{' {one_decl | a_stmnt}+ '}'
a_stmnt     ::= <port_name> '?' {<var_name>|const}+
              | <port_name> '!' {a_expr}+
              | <label_name> ':' a_stmnt
              | 'goto' <label_name>
              | <var_name> '=' aexpr
              | 'if' options 'fi'
              | 'do' options 'od'
links       ::= LINK '{' {link}+ '}'
link        ::= port '=>' {port}+
port        ::= <port_name> IN <autom_name>
```

The body of an automaton is defined as `a_body`, which is the same as `body` in Promela except that `a_body` only allows (for the time being) a subset of the rules of Promela (listed in the rule for `a_stmnt`). We have not shown the rule for `a_expr` but it allows most of the usual Promela expressions, at least for arithmetic and boolean operations.

3.2. The Reactive Promela language. In this section we describe Reactive Promela informally. For the interested reader we have included the formal semantics in appendix A.

3.2.1. *Reactive* Promela *specifications.* A Reactive Promela specification consists of two separate parts:

- a Reactive Promela part consisting of a set of *rproctypes.*
- a Promela part containing any valid Promela code. This is the *pro-active* part of the specification.

3.2.2. *Ports and channels.* The ports in Reactive Promela are *typed* and *directional.* A port is either an *inport* or an *outport.* Ports are declared in the interfaces of rproctypes and automata. As for Promela channels, the types of the message-parameters are declared within braces. The ports declared in the rproctype interface are only references to PROMELA *channels*, declared globally.

3.2.3. *Links.* The link statement is used to create connections between outports and inports. A *valid link* connects exactly one outport to one or more inports. A link statement contains a list of links (separated by a semicolon). Refering to the rule for `link` in the grammar, we see that each link connects an outport (to the left of the `=>` separator) to one or more inports. The port name is qualified with the name of automaton that it belongs to.

3.2.4. *Rproctypes.* An rproctype allows a collection of automata to be encapsulated as a single unit. It consists of:

- A set of *synchronously communicating automata*
- Definition of *links* between the automata.
- Definition of an *interface* (the ports over which the rproctype can communicate with its environment). We can think of this as an *external interface* since it is a subset of interfaces of the automata contained in the rproctype. These ports are only redeclared in the interface—they must be defined as PROMELA *channels* before they can be used in the rproctype.

3.2.5. *Automata.* The REACTIVE PROMELA construct used to specify dynamic behavior is the *automaton.* An automaton definition consists of two parts:

- Definition of an *interface* (the ports over which the automaton can communicate with its environment). Contains the name of each inport and outport together with a list of the types of the arguments that each port can take. Ports that are also declared in the rproctype interface are qualified as `external` ports. For these only the name is declared not the list of argument types.
- Definition of *dynamic behavior.* Given by a directed graph whose nodes are stable- or transitory states and whose edges are transitions labeled by atomic actions. The actions (currently) allowed are those listed under `abody` in the grammar (see section 3.1).

Stable states are the states in which an automaton is waiting for an input on one of its inports. When it receives a message it goes through a reaction phase before ending up in another (possibly the same) stable state. Each automaton has an *initial state* which is stable.

The participation of an automaton to a reaction starts with the execution of the receive statement. The automaton then executes a series of elementary actions (assignments, conditional statements, send statements, ...). External input actions are not allowed during the reaction. The states separating the actions of the reaction are called *transitory states.* The result of such a behavior is a set of outputs to the automaton's environment, as well as an update of the state.

3.2.6. *Reactions.* Now let us explain how the behavior of the automata contained in an rproctype make up the behavior of the rproctype itself. The initial state of an rproctype is a *state vector* containing the initial states of each automaton it encapsulates (with all automata in stable states). When there is a message in one of the channels referred to in the rproctype's interface, the rproctype can go through a *reaction phase* which takes it from one stable state to another (possibly the same).

The reaction starts when one of the encapsulated automata executes an input action to consume a message at the head of a channel. In the sequence of actions that follows one possible action is to send a message to another automata within the same rproctype. This causes the reaction to spread to some or all the automata. An important property of reactions is that all the encapsulated automata are involved in it and that, notably, no external events are taken into account until the reaction terminates. Termination of the reaction is when all the automata again find themselves in a stable state.

Seen from the exterior, the rproctype reacts to an input by changing the state of one or more of the automata it encapsulates, by changing the values of variables local to the automata and by producing a set of messages appended to channels in its interface.

3.3. The modular switch example. To illustrate the utility of REACTIVE PROMELA we will use the example of a simple, modular switch. This is not meant to be a model of a real switch. For a more extensive example, see section 4, where we show how to model the LAPB protocol.

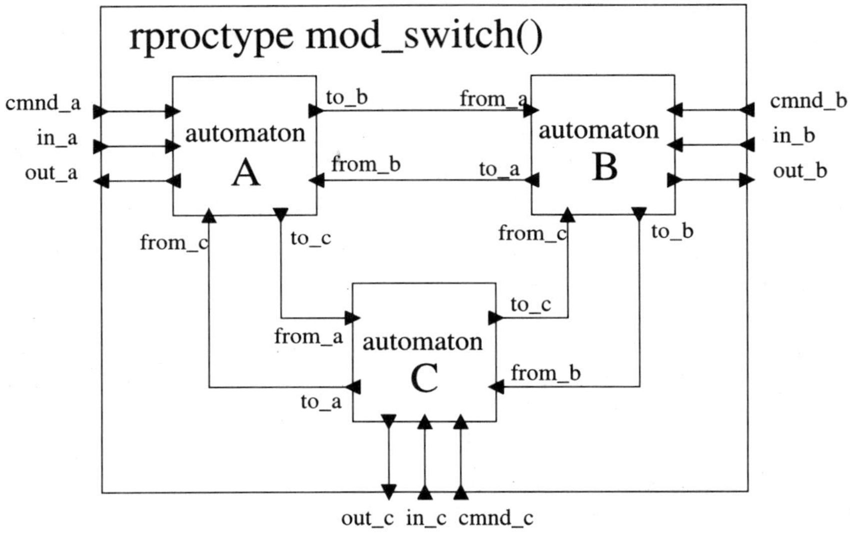

FIGURE 1. Architecture of a reactive switch.

The architecture of the switch is shown in figure 1. It consists of three identical elements, modelled as automata, each of which has three external ports to receive *commands* (on inport `cmnd_x`), and to receive and send packets (on inport `in_x` and on outport `out_x`). Furthermore, each element communicates with each other element through a pair of ports (on outport `to_x` and inport `from_x`). Note that portnames do not have to be unique, since they are declared with the name of the automaton owning it (using the `link` construct). Since each automaton, although fairly big, has a regular structure, it is easier to explain in words what it does than to draw it.

Commands: the configuration of the switch can be changed (by some external manager process) by sending one of the following types of commands:

- *Destroy*—order element to delete all packets it receives.

- *Left*—order element to send all packets to the element on its left.[3]
- *Right*—order element to send all packets to the element on its left.[4]
- *Onward*—order element to output all packets on the output channel.

Each of these commands will take the element to a state whose name corresponds to the command. What is interesting with the reactive style in this example is that an external input is treated atomically. Take the example of element A that receives a packet on port `in_a` while in state left. It will forward the packet to element B. This is still a part of the reaction of the $mod_switch()$ rproctype, hence it in its turn will treat the packet immediately. Lets say that it is in state `onward` meaning that it will send the packet on outport `out_b`. If each of the elements had been separate PROMELA processes, element B might have multiple possibilities: for instance it may have to handle a command or a packet coming in from element C or from the exterior. It is clearly more complicated to design the switch if the switch may be reconfigured while treating a packet.

The simplicity of design permitted by the reactive style also means that it is easy to add new elements to the switch. It suffice to copy an element, rename it, say element D, and change the links.

A final note about this example concerns reactions to multiple inputs. This would be very useful for example to reconfigure all the elements of the switch at the same time. At the moment this is not handled by REACTIVE PROMELA, although we could imagine having an internal manager automaton that would receive a configure message from the exterior with parameters to update all elements of the switch.

3.4. Reactive SPIN. In order to realistically check the correctness of a specification, tool support is essential. Instead of writing a simulator/verifier from scratch we propose to perform a mapping of REACTIVE PROMELA constructs into corresponding ones in PROMELA. For this purpose we present RSPIN, a preprocessor tool that translates a REACTIVE PROMELA specification into an equivalent PROMELA specification.

The translation from a reactive process to a PROMELA process is based on an automata combination algorithm which uses breadth–first search to create the state–space of the automata contained.

One of the most important aims we hope to achieve through the REACTIVE PROMELA extension is to provide a way to reduce the state-space explosion problem. A property of an rproctype is that its reaction to an input from the environment is atomic. This means that many interleavings of actions from the reactive process with actions from other processes need not be calculated during verification. This means that only reactive process actions occuring at stable states are interleaved. In the mapping from REACTIVE PROMELA to PROMELA we take advantage of PROMELA's atomic sequence construct (`atomic { ... }`) to provide a way to implement this property. This also simplifies the use of atomic statements: instead of the user selecting manually which part of a specification to encapsulate in an atomic sequence RSPIN does this automatically.

One difficulty in preserving the semantics of REACTIVE PROMELA in the translation is that PROMELA channels are finite. This means that we cannot be certain that reactions are really atomic. We emphasise that this is not a problem specific

[3] We take left here to mean from element A to B, from B to C and from C to A.

[4] We take right here to mean from element A to C, from B to A and from C to B.

to REACTIVE PROMELA: anyone using the atomic sequence in a PROMELA specification might experience blocking in the middle of a supposedly atomic sequence. One solution is to use SPIN's *"loose messages to full queues"* simulation option. In the RSPIN tool we have also made generation of atomic encapsulation optional to give the user some more flexibility. This option is notably to be used if zero–lenght channels are used to communicate with a reactive process (i.e. rendez–vous communication).

4. Specifying and validating the LAP–B protocol.

In this section we use the specification of the LAP–B protocol[5] to demonstrate the REACTIVE PROMELA language on a nontrivial example. For more information on LAP–B we refer to most textbooks on communications protocols, e.g. [**Sch88**]. In section 4.1 we first show how the protocol is modelled as a REACTIVE PROMELA process through decomposition. Then, in section 4.2 we discuss the translation to PROMELA. Finally, in section 4.3 we give the results from verifying the protocol using a suitable test system.

4.1. Decomposing the LAP–B protocol. To model the LAP–B protocol in REACTIVE PROMELA we decompose it into the five automata[6], playing the following roles:

1. ***Transmitter:*** receives messages from a higher protocol layer and encapsulates each one in a frame to be sent over a physical medium. It sends the frames to the *Retransmitter* process (which will handle possible retransmissions) and the sequence number to the *Window*, and then waits for new messages.
2. ***Window:*** handles the sequence numbers and the sliding window. Once the *Window* is saturated it notifies the *Transmitter* and the upper protocol layer.
3. ***Retransmitter:*** responsible for keeping a buffer of the frames sent but not yet acknowledged. When it receives an acknowledgment number from the *Receiver* it purges all acknowledged frames (in case of reject), and demands the *Acknowledger* process to retransmit the other, non-acknowledged frames.
4. ***Receiver:*** receives frames from the physical medium. Depending on the frame type it decides whether the message is a new one which can be exteacted and sent to the upper protocol layer, whether immediate retransmission is required or not, and so on.
5. ***Acknowledger:*** its role is to (re)transmit one or more frames as indicated by the *Transmitter* or the *Retransmitter*.

This example shows the benefit of decomposition: each automaton that make up the rproctype (i.e. the protocol) have a simple and clearly defined role, but we can still treat and reason about the protocol as a whole. This is because RSPIN translates it into a single PROMELA proctype.

This decomposition arose from an exercise that started with the simplest possible protocol: just a sender and a receiver communicating over a simplex channel.

[5]The example is taken from a course in protocol specification given by Elie Najm at ENST.

[6]This description is not complete, notably with respect to the use of timers; for more information we refer to appendix B which is a complete listing of the REACTIVE PROMELA code for LAP–B.

When extending this to handle duplex communication, finally ending up with the LAP–B protocol, we thought that it might be a good idea to reuse the specifications of the sender and the receiver and combine them into one process. However, combining automata manually is not an easy exercise. If we keep the two processes separate we also noticed that there may be a need for some coordination. An example is that the sender and receiver need to coordinate their actions to decide whether an acknowledgement should be retransmitted as a separate message or to be piggybacked. If both the sender and receiver may receive messages at the same time, coordinating their actions may be difficult. By using the synchronous reactive style of communication we avoid this problem altogether by forbidding other external inputs while one inputs is treated.[7]

4.2. Translating LAP–B **to** PROMELA. Once we have modelled the LAP–B protocol in REACTIVE PROMELA we must use RSPIN translate it to PROMELA before we can perform any simulation and verification. We do not show the complete PROMELA proctype produced, since it is far to big, but we will make a few remarks about it.

We note first that the PROMELA proctype resulting from running RSPIN on the file containing the REACTIVE PROMELA specification of the LAP–B protocol is quite big. The original REACTIVE PROMELA specification is contained in a file of approx. 5 kB (approx. 350 lines of code), whereas the resulting PROMELA proctype is contained in a file of about 200 kB (approx. 5000 lines of code). The resulting proctype has 16 global stable states, which is what we would expect. The full crossproduct of the five automata would give 32 states, but the *Transmitter* and the *Window* have a shared state, corresponding to the case of a saturated window. This shows very clearly that decomposition is essential for modelling complex protocols: we have tried to combine automata manually but gave up after combining the three smallest ones.

Internal communication between automata is reduced to assignment to variables. An example (see appendix B) is that the send statement `F!VS` in the *Transmitter* and the corresponding receive statement in the *Window* `T?VS` is reduced to the assignment `Window_VS=Transmitter_VS`. Since the variable `VS` exists in both the transmitter and in the window it is prefixed with the automaton name in the combined automaton. A more interesting case is where the *Window* notifies both the *Transmitter* and the upper protocol layer that it is saturated. This is an example of a communication with more than one receiver. Between the transmitter and the window this is a pure synchronization which takes the combined automaton to stable state `s1` (the saturation state). But we still need to notify the upper layer: therefore the combined automaton keeps the send action `CF!Xoff`. All external communication actions are kept as is in the combined automaton.

4.3. Simulating and verifying the LAP–B **protocol.** To check that the specification of the LAP–B presented above is correct we use SPIN to simulate and verify the translated protocol. Note that in this section we only talk about normal PROMELA processes.

We use a standard system to test the protocol: one process models a communication medium which non-deterministically looses, corrupts or transfers correctly

[7]Note that in many of the other syncronous languages, e.g. Esterel [**BG91**] or SL [**BdS95**], processes react to a collection of inputs. We consider making this possible in REACTIVE PROMELA as well.

a frame; on each side of the link is an instance of the LAP–B process. Each LAP–B instance communicates with a simple user process (which sends a finite number of messages using the protocol), and with two timer processes—one for retransmissions if no acknowledgement arrives in time[8] and one for keeping track of how long to wait before acknowledging a message with a separate frame instead of waiting for a message to piggyback the acknowledgement on.

We can now use SPIN to analyse the protocol. By random or step-by-step simulation we have checked that the test system in the basic cases behaves as expected. However, when we used the exhaustive simulation (verification) mode, we detected an error in *our specification of* the LAP–B protocol. The problem occurs when the *Windows* on both sides of the communication link are saturated. Saturation means that acknowledgements will not be piggybacked. If a frame is received correctly by the saturated side and it sends an acknowledgement which is lost, it will not retransmit the acknowledgement when the other side retransmits the frame. The other side retransmits after a timeout period, but the acknowledgements are not retransmitted (unless they can be piggybacked).

It would of course be interesting to compare the size of the graph created during verification of our system with the graph for a corresponding system where everything is modelled as PROMELA processes. Unfortunately we do not have a machine with enough memory available to achieve this.

In addition to the LAP–B protocol we have also modelled the **Go–Back–N** and the **Selective Reject** protocols. So far we have only simulated them, but we will soon try to verify these protocols as well.

5. Conclusion

In this paper we have presented the REACTIVE PROMELA language and its associated tool RSPIN through the specification of the data link protocol LAP–B. The language belongs to the family of synchronous reactive formalisms and allows a system to be decomposed into a reactive part containing configurations of synchronously communicating automata and a pro-active part containing PROMELA proctypes.

The RSPIN tool translates rproctypes into PROMELA proctypes, so that a REACTIVE PROMELA specification can be simulated and verified with SPIN. No modifications to the SPIN tool are needed to do this. We saw that by translating a REACTIVE PROMELA specification to a PROMELA specifications using RSPIN, we could simulate and verify it using SPIN.

References

[AF90] C. André and L. Fancelli, *A mixed (asynchronous / synchronous) implementation of a real-time system*, Euromicro 90, Amsterdam, 1990.

[BB91] G. Berry and A. Benveniste, *The synchronous approach to reactive and real-time systems*, Another Look at Real Time Programming, Proceedings of the IEEE **79** (1991), 1270–1282.

[BCGH93] Albert Benveniste, Paul Caspi, Paul Le Guernic, and Nicolas Halbwachs, *Data-flow synchronous languages*, Rapport de recherche 2089, INRIA, Unité de recherche INRIA Sophia-Antipolis, France., October 1993.

[8]Of course, PROMELA doesn't model time—a timeout can only occur if no other action is possible—which makes it difficult to model this accurately. Therefore proposed timed extensions of PROMELA [**TC96**] are very interesting.

[BdS95] Frédéric Boussinot and Robert de Simone, *The sl synchronous language*, Rapport de recherche 2510, INRIA, Unité de recherche INRIA Sophia-Antipolis, France., Mars 1995.

[Ber93a] G. Berry, *Communicating reactive processes*, Proc. 20th ACM Conf. on Principles of Programming Languages, Charleston, Virginia, 1993.

[Ber93b] G. Berry, *The semantics of pure esterel*, Proc Marktoberdorf Intl. Summer School on Program Design Calculi, LNCS, to appear, Springer-Verlag, 1993.

[BG91] G. Berry and G. Gonthier, *Incremental development of an hdlc entity in Esterel*, Comp. Networks and ISDN Systems **22** (1991), 35–49.

[BG92] G. Berry and G. Gonthier, *The Esterel synchronous programming language: Design, semantics, implementation*, Science Of Computer Programming **19** (1992), no. 2, 87–152.

[BN83] S. Budkowski and E. Najm, *Structured finite state automata. a new approach for modelling distributed communication systems.*, Protocol Specification, Testing and Verification, III (H. Rudin and C. H. West, eds.), Elsevier Science Publishers B.V (North-Holland), 1983.

[Bou91] F. Boussinot, *Reactive c: An extension of c to program reactive systems*, Software-Practice and Experience **21** (1991), no. 4, 401–428.

[CPHP85] P. Caspi, D. Pilaud, N. Halbwachs, and J. Plaice, *Lustre, a declarative language for programming synchronous systems*, Proceedings of ACM Conference on Principles of Programming Languages, ACM, 1985.

[dSdS95] Monica Lara de Souza and Robert de Simone, *Using po methods for verifying behavioural equivalences*, Proceedings of FORTE'95, October 1995, pp. 59–74.

[Fer89] Jean-Claude Fernandez, *Aldebaran: A tool for verification of communicating processes*, Rapport SPECTRE C14, Laboratoire de Génie Informatique — Institut IMAG, Grenoble, September 1989.

[GKPP95] R. Gerth, R. Kuiper, R. Peled, and W. Penczek, *A partial order approach to branching time model checking*, Proceedings of ISTCS, 1995, pp. 330–339.

[God94] Patrice Godefroid, *Partial-order methods for the verification of concurrent systems: An approach to the state-explosion problem*, Ph.D. thesis, UNIVERSITE DE LIEGE, Faculté des Sciences Appliquées, 1994.

[Gue86] P. Le Guernic, *Signal, a data-flow oriented language for signal processing*, IEEE Trans. ASSP **34** (1986), no. 2, 362–374.

[Hal93] N. Halbswachs, *Synchronous programming of reactive systems*, Kluwer Academic Press, Netherlands, 1993.

[Hol91] Gerhard Holzmann, *Design and validation of computer protocols*, first ed., Prentice-Hall, Englewood Cliffs, N.J., 1991.

[JLRM] M. Jourdan, F. Lagnier, P. Raymond, and F. Maraninchi, *A multiparadigm language for reactive systems.*

[Mad92] E. Madelaine, *Verification tools from the Concur project*, EATCS Bulletin **47** (1992).

[MV89] E. Madelaine and D. Vergamini, *Auto: A verification tool for distributed systems using reduction of finite automata networks*, Proc. FORTE'89 Conference, Vancouver, 1989.

[NO96] Elie Najm and Frank Olsen, *Reactive efsms - reactive promela/rspin*, Tools and Algorithms for the Construction and Analysis of Systems: second international workshop ; proceedings / TACAS'96, Passau, Germany, March 27–29 (Berlin) (Tiziana Margaria and Bernhard Steffen, eds.), Lecture Notes in Computer Science, vol. 1055, Springer-Verlag, 1996, pp. 349–368.

[Pel94] D. Peled, *Combining partial order reductions with on-the-fly model-checking*, Proceedings of CAV'94, LNCS 818, Springer-Verlag, 1994.

[Plo81] G. Plotkin, *A structural approach to operational semantics*, Tech. report, Comput. Sci. Dept., Aarhus Univ., 1981.

[RdS90] V. Roy and R. de Simone, *Auto and autograph*, proceedings of Workshop on Computer Aided Verification (New-Brunswick) (R. Kurshan, ed.), AMS-DIMACS, June 1990.

[Sch88] Mischa Schwarz, *Telecommunications networks: Protocols, modeling and analysis*, Addison–Wesley Series in Electrical and Computer Engineering, Addison-Wesley, Reading, MA, USA, 1988.

[TC96] Stavris Tripakis and Costas Courcoubetis, *Extending promela and spin for real time*, Tools and Algorithms for the Construction and Analysis of Systems: second international workshop ; proceedings / TACAS'96, Passau, Germany, March 27–29 (Berlin) (Tiziana Margaria and Bernhard Steffen, eds.), Lecture Notes in Computer Science, vol. 1055, Springer-Verlag, 1996, pp. 329–348.

[Val90] A. Valmari, *A stubborn attack on state explosion*, LNCS 531, Springer-Verlag, 1990.

[WG93] P. Wolper and P. Godefroid, *Partial order methods for temporal verification*, Proceedings of Concur'93, LNCS 715, Springer-Verlag, 1993.

Appendix A. Semantics of Reactive PROMELA

In order to reason about our extension and about the translation from REACTIVE PROMELA to PROMELA we formalise an essential subset of PROMELA in section A.1, in terms of PROMELA *State Machines (PSM)*. In section A.2 we then formalise the reactive extension with the *Reactive State Machine (RSM)* model.[9]

A.1. PROMELA State Machines.

The subset of PROMELA that consider can be formally defined using the following settings. We consider first a set L of elementary PROMELA instructions (with typical elements l):

$$L ::= [pred] \mid v = Exp \mid c!Exp \mid c?v$$

These instructions correspond to test, assignment, send, and receive instructions. The send and receive instructions are to bounded channels c. We consider simple channels containing FIFO sequences of simple values.

A.1.1. *PSM processes.* We consider PSM processes (with generic element P) as follows: P is a tuple (S, s, T, E) where:

- S is a set of *states*;
- s is either the *initial state* or the *current state*;
- $T \subset S \times L \times S$ is a *transition relation*;
- E is the *environment* of variables in P. It consists of a set of (untyped) variables V (ranging over values in the set VAL) and a mapping of each variable $v \in V$ to a value $val \in VAL$.

We use a dot-notation to access elements of the tuple representing P. $P.T$ is the transition relation T of PSM process P and $P.E$ is the environment of P.

A.1.2. *Semantics of PSM processes.* We give the semantics of a PSM process P by a translation function from PSM into a *Labeled Transition System (LTS)* defined by: $(PSM \times \Gamma \times PSM)$.[10] A PSM transition (P, α, P'), also written $P \xrightarrow{\alpha} P'$, means that P can perform action α to become P'.

The following notations are used in the rules:

- $E \vdash Exp \rightarrow val$ denotes that the expression Exp evaluates to val in the environment E.
- $E \oplus (v : val)$ denotes an environment E obtained from E by updating variable v to val.

The translation from PSM to LTS is given by the following set of SOS rules:

[9] Note that this appendix is taken from our paper presented at TACAS'96 [**NO96**] It is included here for completeness and in order to resolve any ambiguities in our informal, but more intuitive description in section 3.

[10] where $\Gamma ::= \epsilon \mid c!val \mid c?val$, with a typical element α.

TEST

$$\frac{(s,[pred],s') \in T \quad E \vdash pred \rightarrow True}{(S,s,T,E) \xrightarrow{\epsilon} (S,s',T,E)}$$

ASSIGNMENT

$$\frac{(s,v = Exp,s') \in T \quad E \vdash Exp \rightarrow val}{(S,s,T,E) \xrightarrow{\epsilon} (S,s',T,E \oplus (v : val))}$$

SEND

$$\frac{(s,c!Exp,s') \in T \quad E \vdash Exp \rightarrow val}{(S,s,T,E) \xrightarrow{c!val} (S,s',T,E)}$$

RECEIVE

$$\frac{(s,c?v,s') \in T}{(S,s,T,E) \xrightarrow{c?val} (S,s',T,E \oplus (v : val))}$$

A.1.3. *Atomic PSM processes.* We now consider a larger subset of PROMELA containing the *atomic*{...} construct. Consequently, our model is extended to reflect this construct. We define atomic PSM processes[11] (with typical element Q) as a triple (P,Π,σ) where:

- P is a PSM process.
- Π is a *partioning* of $P.S$ into a set of disjoint non-empty sets of states, i.e. $\forall p_1,p_2 \in \Pi : p_1 \cap p_2 = \emptyset$ where $p_1 \in P.S, p_2 \in P.S$. Note that there may be some states in $P.S$ that are not in any partition $p \in \Pi$.
- $\sigma = \downarrow \emptyset \mid \downarrow p \mid \uparrow p$ is the current *atomic section* of P. $\downarrow \emptyset$ denotes that P is not in an atomic section, $\downarrow p$ denotes that P has entered atomic section p but is not yet active, and $\uparrow p$ denotes that P is active (executing a sequence of atomic steps) in p.

A.1.4. *Semantics of atomic PSM processes.* The semantics of atomic PSM processes is given by two rules. We use the function $\Pi(s)$ defined by $\Pi(s) = p \in \Pi$ (for $s \in p$) and $\Pi(s) = \emptyset$ (for $s \notin \bigcup_{p\in\Pi} p$).

DEACTIVATED–ATOMIC–SEQUENCE

$$\frac{P \xrightarrow{\alpha} P' \quad (\Pi(P'.s) \neq p \ \vee\ \Pi(P'.s) = \emptyset)}{\begin{array}{c}(P,\Pi,\downarrow p) \xrightarrow{\alpha} (P',\Pi,\downarrow \Pi(P'.s)) \\ (P,\Pi,\uparrow p) \xrightarrow{\alpha} (P',\Pi,\downarrow \Pi(P'.s))\end{array}}$$

ACTIVATED–ATOMIC–SEQUENCE

[11] we do allow for non-atomic PSM processes as a special case where $Q = (P,\emptyset,\downarrow \emptyset)$.

$$\frac{P \xrightarrow{\alpha} P' \quad p \neq \emptyset \quad \Pi(P'.s) = p}{\begin{array}{c}(P, \Pi, \downarrow p) \xrightarrow{\alpha} (P', \Pi, \uparrow p) \\ (P, \Pi, \uparrow p) \xrightarrow{\alpha} (P', \Pi, \uparrow p)\end{array}}$$

Figure 2 illustrates the rules and shows the different cases for activation and deactivation of atomic sequences. The transitions numbered (1) to (6) correespond to the first rule (DEACTIVATED–ATOMIC–SEQUENCE), whereas transitions (7) and (8) correspond to the second rule (ACTIVATED–ATOMIC–SEQUENCE).

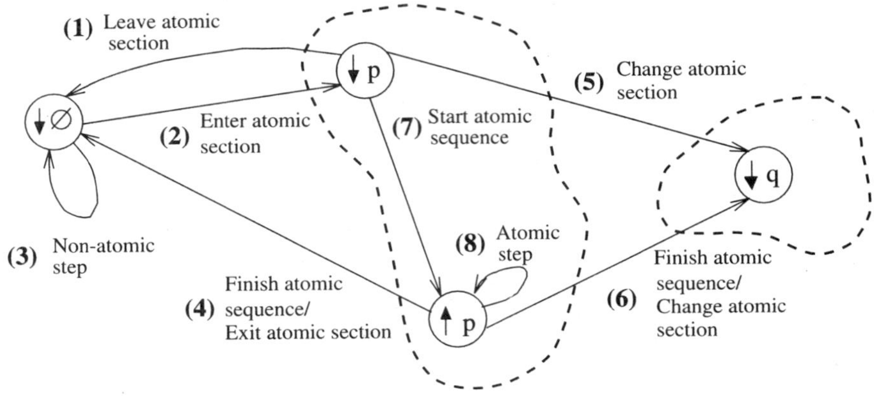

FIGURE 2. Allowable atomic transitions.

A.1.5. *PSM specifications.* We now turn to complete PSMs. A PSM specification is a pair $(Procs, Chans)$ where:

- $Procs$ is a set of atomic PSM processes;
- $Chans$ is a set of bounded FIFO channels. If the length of the channel is zero then communication is by rendez–vous, else it is by asynchronous message passing.

A.1.6. *Semantics of PSM specifications.* We use the following notations in the rules: $Int(Q)$ means that Q is *interruptible* (i.e. for $Q = (P, \Pi, \sigma) : \sigma = \downarrow \emptyset$), $\#c$ gives the current number of values (messages) in channel c, $l(c)$ gives the length of channel c, and $head(c)$ gives the value at the head of the channel c.

SPEC–ASYNCH–SEND

$$\frac{Q_i \xrightarrow{c!v} Q_i' \quad \forall j \neq i : Int(Q_j) \quad \#c < l(c) \quad l(c) \neq 0}{(Chans, Procs) \longrightarrow (Chans', Procs')}$$

$$\text{where} \quad \begin{array}{l} Procs' = (Procs - \{Procs'\}) \cup Q_i' \\ Chans' = Chans - \{c(v_1, \ldots, v_n)\} \cup \{c(v_1, \ldots, v_n, v)\} \end{array}$$

SPEC–ASYNCH–RECEIVE

$$\frac{Q_i \xrightarrow{c?v} Q_i' \quad \forall j \neq i : Int(Q_j) \quad \#c > 0 \quad head(c) = v}{(Chans, Procs) \longrightarrow (Chans', Procs')}$$

$$\text{where} \quad \begin{array}{l} Procs' = (Procs - \{Procs'\}) \cup Q_i' \\ Chans' = Chans - \{c(v_1, \ldots, v_n, v)\} \cup \{c(v_1, \ldots, v_n)\} \end{array}$$

SPEC–RENDEZ-VOUS

$$\frac{Q_i \xrightarrow{c!v} Q_i' \quad Q_j \xrightarrow{c?v} Q_j' \quad \forall k \neq i : Int(Q_k) \quad l(c) = 0}{(Chans, Procs) \longrightarrow (Chans, Procs')}$$

$$\text{where } Procs' = (Procs - \{Procs'\}) \cup Q_i' \cup Q_j'$$

SPEC–INTERNAL

$$\frac{Q_i \xrightarrow{\epsilon} Q_i' \quad \forall j \neq i : Int(Q_j)}{(Chans, Procs) \longrightarrow (Chans, Procs')}$$

$$\text{where } Procs' = (Procs - \{Procs'\}) \cup Q_i'$$

A.2. Reactive State Machines. Here we present the *Reactive State Machine (RSM)*, a formalisation of our proposed extension to PROMELA.

Like we did for PSMs, we give the syntax and semantics in an incremental fashion, starting with *RSM automata*, then *RSM processes*, and finally we give the semantics of complete *RSM specifications*. The set of instructions L is the same as the set used in section A.1.

A.2.1. *RSM automata.* In the RSM model a process can be decomposed into a configuration of one or more *RSM automata*. An RSM automaton, A, is a six–tuple $(S, s, spred, T, E, I)$ where:

- S is a set of states partioned into two disjoint subsets[12]:
  - $SS \subseteq S$ a set of *stable states*
  - $TS \subset S$ a set of *transitory states*
- s is either the *initial state* $\in SS$ or the *current state.*
- *spred* : $s \longrightarrow \{True, False\}$ is a function to determine if a give state is stable or transitory. $spred(s) = True$ if $s \in SS$ and $spred(s) = False$ if $s \in TS$.
- $T \subset S \times L \times S$ is a *transition relation*
- E is the *environment* of variables in A.
- I is an *interface*[13] consisting of:
  - P_{in} a set of *inports*
  - P_{out} a set of *outports*

A *well-formed* RSM automaton has the following restrictions:

[12]where $S = SS \cup ST$ and $SS \cap ST = \emptyset$

[13]where $P_{in} \cap P_{out} = \emptyset$.

- if $(s, l, s') \in T$ and $s \in SS$ then l is an input action.
- if $(s, l, s') \in T$ and $l = c?x$ then $c \in P_{in}$.
- if $(s, l, s') \in T$ and $l = c!val$ then $c \in P_{out}$.

A.2.2. *Semantics of RSM automata.* The semantics of an RSM automaton is given by the same rules as for a PSM process.

A.2.3. *RSM processes.* An RSM process R is a tuple (A, I, L) where:

- A is a set of RSM automata $\{a_1, \ldots, a_n\}$
- I is an *interface*[14] consisting of:
  - C_{in} a set of *input channels*
  - C_{out} a set of *output channels*
- L is a set of *links* taking the following three forms:
  - L_{int} a set of *internal links* given by the tuple $((a_i, p_i), (a_j, p_j))$, where $p_i \in a_i.P_{out}$ and $p_j \in a_j.P_{in}$.
  - L_{out} a set of *external output links* given by the tuple $((a_i, p_i), (\bullet, p_j))$, where $p_i \in a_i.P_{out}$ and $p_j \in C_{out}$.
  - L_{in} a set of *external input links* given by the tuple $((\bullet, p_i), (a_j, p_j))$, where $p_i \in C_{in}$ and $p_j \in\in a_j.P_{in}$.

Some other definitions:

- the *state* of an RSM process R is the tuple $R.s = (a_1.s, \ldots, a_n.s)$, where:
  - the *initial state* is defined as the tuple $(a_1.s_0, \ldots, a_n.s_0)$
  - a state $s = (s_1, \ldots, s_n)$ is a *stable state* iff $\forall s_i : spred_i(s_i) = True$.
  - a state $s = (s_1, \ldots, s_n)$ is a *transitory state* iff $\exists s_i : spred_i(s_i) = False$.

A.2.4. *Semantics of RSM processes.* We add the predicate $stable(R) = R.s$, meaning that RSM process R is in a stable state, and give the SOS rules for RSM processes:

RSM–PROC–EXT-SEND

$$\frac{a_i \xrightarrow{g!val} a_i' \quad ((a_i, g), (\bullet, c)) \in L_{out} \quad c \in C_{out} \quad l(c) < \#c \quad \#c \neq 0}{(A, I, L) \xrightarrow{c!val} (A', I, L)}$$

where $A = a_1, \ldots, a_i, \ldots, a_n$, $A' = a_1, \ldots, a_i', \ldots, a_n$

RSM–PROC–EXT-RECEIVE

$$\frac{\begin{array}{l} a_i \xrightarrow{g?v} a_i' \\ ((\bullet, c), (a_i, g)) \in L_{in} \quad c \in C_{in} \quad l(c) \neq 0 \quad \#c \neq o \quad stable((A, I, L)) \end{array}}{(A, I, L) \xrightarrow{c?v} (A', I, L)}$$

where $A = a_1, \ldots, a_i, \ldots, a_n$, $A' = a_1, \ldots, a_i', \ldots, a_n$

[14]where $C_{in} \cap C_{out} = \emptyset$

RSM–PROC–SYNCH

$$\frac{a_i \xrightarrow{g!val} a_i' \quad a_j \xrightarrow{h?v} a_j' \quad ((a_i,h),(a_j,g)) \in L_{int} \quad \#c = 0}{(A,I,L) \xrightarrow{\epsilon} (A',I,L)}$$

$$\text{where} \quad \begin{array}{l} A = a_1,\ldots,a_i,\ldots,a_j,\ldots,a_n \\ A' = a_1,\ldots,a_i',\ldots,a_j',\ldots,a_n \end{array}$$

A synchronisation between automata a_i and a_j is possible only if a_I can send a message on outport p, a_I can receive a message on inport q, and there is a link between (a_i, p) and (a_j, q).

RSM–PROC–INTERNAL–ACTION

$$\frac{a_i \xrightarrow{\epsilon} a_i'}{(A,I,L) \xrightarrow{\epsilon} (A',I,L)}$$

$$\text{where} \quad \begin{array}{l} A = a_1,\ldots,a_i,\ldots,a_n \\ A' = a_1,\ldots,a_i',\ldots,a_n \end{array}$$

A.2.5. *RSM specifications.* An RSM specification can be given by a the triple $(RProcs, Procs, Chans)$ where:

- $RProcs$ is a set of RSM processes;
- $Procs$ is a set of atomic PSM processes;
- $Chans$ is a set of channels.

A.2.6. *Semantics of RSM specifications.* The semantics of RSM specifications can simply be given using the rules for PSM specifications where the predicate $Int(R)$ is defined on RSM process R by $Int(R) = stable(R)$.

Appendix B. Reactive PROMELA code for LAP–B protocol

Below follows a listing of the REACTIVE PROMELA specification of the LAP–B data link protocol. We do not show the code generated by RSPIN as it is far to big (approx. 5000 lines, as opposed to approx. 350 for the REACTIVE PROMELA specification).

```
#define QSIZE       10
#define Xoff        0
#define Xon         1
#define armer       1
#define desarmer    0
#define sonnerie    2
#define N           0
#define U           1
#define I           2
#define RR          3
#define REJ         4
#define cor         5
```

```
#define f           2
#define W           4

typedef Frame {
  byte type;
  byte D;
  byte NS;
  byte NR
}

/* for the inports */
chan UE[2]  = [QSIZE] of { int };
```

```
chan LR[2]  = [QSIZE] of { byte,byte,
                           byte,byte  };
chan T1r[2] = [QSIZE] of { byte };
chan T2r[2] = [QSIZE] of { byte };
/* for the outports */
chan UR[2]  = [QSIZE] of { int };
chan CF[2]  = [QSIZE] of { byte };
chan T1d[2] = [QSIZE] of { byte };
chan T2d[2] = [QSIZE] of { byte };
chan LE[2]  = [QSIZE] of { byte,byte,
                           byte,byte  };

rproctype HDLC
  (inport  UE, T1r, T2r, LR;
   outport UR, CF, T1d, T2d, LE)
  (bit no)
{

  automaton Tra
    (external inport UE;
     inport   CF={bool};
     outport  RA={byte,byte,byte,byte},
              F={byte})
    (/* no fpars */)
  {
    byte VS=1, M;
    Frame F;

  tra0:
    if
    :: UE[no]?M ->
       F.type=I;
       F.D=M;
       F.NS=VS;
       VS=(VS+1)%W;
       RA!F.type,F.D,F.NS,F.NR;
       F!VS;
       goto tra0
    :: CF[no]?Xoff ->
       goto tra1
    fi;

  tra1:
    CF[no]?Xon ->
    goto tra0
  }

  automaton Fen
    (external outport CF={bool};
     inport   T={byte}, C={byte})
    (/* no fpars */)
  {
    byte VS, VA, va;
    bool x, y;

  fen0:
    if
    :: T?VS ->
       x = ((VS-VA>0 && VS-VA<=f) ||
            (VS-VA+W>0 && VS-VA+W<=f));
       if
       :: (!x) ->
          CF[no]!Xoff;
          goto fen1
       :: (x) ->
          goto fen0
       fi
    :: C?va ->
       y = ((va-VA>0 && va-VA<=f) ||
            (va-VA+W<=f && va-VA+W>0));
       if
       :: (!y) ->
          goto fen0
       :: (y)  ->
          VA=va;
          goto fen0
       fi
    fi;

  fen1:
    C?va ->
    y = ((va-VA>0 && va-VA<=f) ||
         (va-VA+W<=f && va-VA+W>0));
    if
    :: (!y) ->
       goto fen1
    :: (y)  ->
       VA=va;
       CF[no]!Xon;
       goto fen0
    fi
  }

  automaton Ret
    (external inport  T1r;
     external outport T1d;
     inport   C={byte,byte},
              T={byte,byte,byte,byte};
     outport  A={byte,byte,byte,byte})
    (/* no fpars */)
  {
    byte RT, ni, n, nf, i, j, b;
    Frame Buf[f];

  ret0:
    if
    :: T?Buf[ni].type,
         Buf[ni].D,
         Buf[ni].NS,
         Buf[ni].NR ->
       n=1;
       T1d[no]!armer;
       goto ret1
    :: C?b,RT ->
       goto ret0
    fi;
```

```
ret1:
  if
  :: T?Buf[(ni+n)%f].type,
       Buf[(ni+n)%f].D,
       Buf[(ni+n)%f].NS,
       Buf[(ni+n)%f].NR ->
     nf=(ni+n)%f;
     n++;
     goto ret1
  :: C?b,RT ->
     i=0;
     j=n;
     do
     :: (i == j) ->
        n=j;
        break
     :: (i != j) ->
        if
        :: (Buf[(ni+i)%f].NS == RT) ->
           ni=(ni+i+1)%f;
           n--;
           break
        :: (Buf[(ni+i)%f].NS != RT) ->
           i++;
           n--
        fi
     od;
     if
     :: (n == 0) ->
        T1d[no]!desarmer;
        goto ret0
     :: (n != 0) ->
        if
        :: b == N ->
           goto ret1
        :: b == U ->
           i=0;
           do
           :: (i != n) ->
              A!Buf[(ni+i)%f].type,
                Buf[(ni+i)%f].D,
                Buf[(ni+i)%f].NS,
                Buf[(ni+i)%f].NR;
              i++
           :: (i == n) ->
               T1d[no]!armer;
              goto ret1
           od
        fi
     fi
  :: T1r[no]?sonnerie ->
     i=0;
     do
     :: (i != n) ->
        A!Buf[(ni+i)%f].type,
          Buf[(ni+i)%f].D,
          Buf[(ni+i)%f].NS,
          Buf[(ni+i)%f].NR;
        i++
     :: (i == n) ->
        T1d[no]!armer;
        goto ret1
     od
  fi
}

automaton Rec
  (external inport  LR;
   external outport UR;
   outport  A={byte,byte},
            R={byte,byte},
            F={byte})
  (/* no fpars */)
{
  Frame F;
  byte VR=1, i;
  bool x;

rec0:
  LR[no]?F.type,F.D,F.NS,F.NR ->
  if
  :: (F.type == cor) ->
     goto rec0
  :: (F.type != cor) ->
     F!F.NR;
     if
     :: (F.type == RR) ->
        R!N,F.NR;
        goto rec0
     :: (F.type == REJ) ->
        R!U,F.NR;
        goto rec0
     :: (F.type == I) ->
        R!N,F.NR;
        x =
        ((F.NS-VR>0 && F.NS-VR<f) ||
        (F.NS-VR+W<f && F.NS-VR+W>0));
        if
        :: (!x && F.NS != VR) ->
           i++;
           if
           :: (i != f) ->
              goto rec0
           :: (i == f) ->
              A!U,(VR+W-1)%W;
              i=0;
              goto rec0
           fi
        :: (F.NS == VR) ->
           UR[no]!F.D;
           A!N,VR;
           VR=(VR+1)%W;
           i=0;
           goto rec0
        :: (x) ->
```

```
            A!U,VR;
            goto rec1
          fi
       fi
    fi;

rec1:
  LR[no]?F.type,F.D,F.NS,F.NR ->
    if
    :: (F.type == cor) ->
       goto rec1
    :: (F.type != cor) ->
       F!F.NR;
       if
       :: (F.type == RR) ->
          R!N,F.NR;
          goto rec1
       :: (F.type == REJ) ->
          R!U,F.NR;
          goto rec1
       :: (F.type == I) ->
          R!N,F.NR;
          if
          :: (F.NS != VR) ->
             goto rec1
          :: (F.NS == VR) ->
             UR[no]!F.D;
             A!N,VR;
             VR=(VR+1)%W;
             i=0;
             goto rec0
          fi
       fi
    fi
}

automaton Ack
  (external inport  T2r;
   external outport T2d, LE;
   inport   C={byte,byte},
            T={byte,byte,byte,byte},
            R={byte,byte,byte,byte})
  (/* no fpars */)
{
  Frame F;
  byte b,NA;

ack0:
  if
  :: T?F.type,F.D,F.NS,F.NR ->
     F.NR=NA;
     LE[no]!F.type,F.D,F.NS,F.NR;
     goto ack0
  :: R?F.type,F.D,F.NS,F.NR ->
     F.NR=NA;
     LE[no]!F.type,F.D,F.NS,F.NR;
     goto ack0
  :: C?b,NA ->
     if
     :: (b == N) ->
        T2d[no]!armer;
        goto ack1
     :: (b == U) ->
        F.type=REJ;
        F.NR=NA;
        LE[no]!F.type,F.D,F.NS,F.NR;
        goto ack0
     fi
  fi;

ack1:
  if
  :: T?F.type,F.D,F.NS,F.NR ->
     T2d[no]!desarmer;
     F.NR=NA;
     LE[no]!F.type,F.D,F.NS,F.NR;
     goto ack0
  :: R?F.type,F.D,F.NS,F.NR ->
     T2d[no]!desarmer;
     F.NR=NA;
     LE[no]!F.type,F.D,F.NS,F.NR;
     goto ack0
  :: C?b,NA ->
     if
     :: (b == N) ->
        goto ack1
     :: (b == U) ->
        T2d[no]!desarmer;
        F.type=REJ;
        F.NR=NA;
        LE[no]!F.type,F.D,F.NS,F.NR;
        goto ack0
     fi
  :: T2r[no]?sonnerie ->
     F.type=RR;
     F.NR=NA;
     LE[no]!F.type,F.D,F.NS,F.NR;
     goto ack0
  fi
}

link {
  A  in Ret => R  in Ack;
  F  in Tra => T  in Fen;
  RA in Tra => T  in Ret, T in Ack;
  CF in Fen => CF in Tra;
  F  in Rec => C  in Fen;
  R  in Rec => C  in Ret;
  A  in Rec => C  in Ack
}
}
```

Networks Department, Ecole Nationale Supérieure des Télécommunications, 46, rue Barrault, 75013 Paris, France

E-mail address: najm@res.enst.fr

Networks Department, Ecole Nationale Supérieure des Télécommunications, 46, rue Barrault, 75013 Paris, France

Current address: Centre National d'Etudes des Télécommunications, 38-40, rue du Général Leclerc, 92131 Issy-Les-Moulineaux, France

E-mail address: frank.olsen@issy.cnet.fr

DIMACS Series in Discrete Mathematics
and Theoretical Computer Science
Volume **32**, 1997

Outline for an Operational Semantics of PROMELA

V. Natarajan and Gerard J. Holzmann

ABSTRACT. PROMELA is a high-level specification language for modeling interactions in distributed systems, and for expressing logical correctness requirements about such interactions. The model checker SPIN accepts specifications written in this language, and it can produce automated proofs for each type of property. SPIN either proves that a property is valid in the given system, or it generates a counter-example that shows that it is not. This paper contains the outline for an operational-semantics definition of PROMELA.

1. Introduction

PROMELA is a language for reasoning about concurrent systems. The language has developed over the last fifteen years from a simple modeling language into a more complete specification language. The language, supported by the model checker SPIN, is used in a growing number of design projects, both large and small.

As new constructs are introduced into the language (e.g., [**Hol95**]), their interplay with existing constructs can have subtle consequences. Some combinations of constructs can reasonably be interpreted in more then one way. So far, only the implementation of the model checker SPIN can provide definitive answers about semantics issues. A formal semantics definition can provide an implementation independent way to resolve possible conflicts of interpretation.

This paper gives an outline for a semantics basis of the language. To simplify the exposition, it imposes a number of restrictions. It will not discuss, for instance, *array variables* or *structures*, nor will it discuss *assert* statements, *never claims*, or correctness properties expressed by *progress* and *accept* labels. To simplify the definitions somewhat, we will also restrict *run statements* to one single parameter, and *channel declarations* to a single message field.

1991 *Mathematics Subject Classification.* Primary 68Q65, 68Q22, 68M15; Secondary 68N20.

Key words and phrases. Verification, Semantics, PROMELA, SPIN.

Work done during the first author's stay at Bell Laboratories in the summer of 1995.

The remainder of this paper is organized as follows. Section 2 introduces notational conventions, and some semantic preliminaries. Section 3 formalizes the notion of *executability* of PROMELA statements. Section 4 presents how global states are transformed by the execution of statements. The notion of the execution of a program is captured in Section 5. Section 6 gives a small example of the conversion of a program into a labeled transition system. Section 7 summarizes the report.

2. Preliminaries

A PROMELA program consists of zero or more global variable declarations, and one or more process type declarations (*proctypes*). The verifier SPIN translates each process type declaration into a labeled transition system, in which the labels are primitive statements [**Hol94**]. We define below what a *primitive statement* is. For the time being it will suffice to note that a primitive statement is a state transformer: when it is *executable* (to be defined) it modifies the *current state* (to be defined) in a precise way (to be defined). Note that compound statements (such as *if .. fi*, *do .. od*, *unless* (introduced in [**Hol95**]), and *goto* are not primitive statements, and neither are *else* statements or declarations.

Converting compound statements, or generally process type declarations, into labeled transition systems is a fairly straightforward procedure that is not detailed here. (It will be detailed in the full version.) Labeled transition systems are formally defined in Section 2.4. An example of the translation from a *proctype* declaration into a labeled transition system is given in Section 6. The conversion is performed by SPIN when it parses its input. The result can be seen after a verifier is generated and compiled, with: `pan -d`. (A graphical representation can also be seen with the *FSM View* option in XSPIN.)

2.1. Notational Conventions. If S is a nonempty set then S^i will denote the set of sequences of length i whose elements belong to S. If x is a sequence, then $x.i$ denotes the ith element of x. We will often assign symbolic names to specific positions in a sequence; if *sym* is the symbolic name for the ith position of the sequence x then $x.sym$ will denote $x.i$. If x is any mathematical structure then $x[y/sym]$ will stand for the same structure in which the substructure denoted by *sym* is replaced by y. If f is a function then $f[sym = y]$ will denote the function which takes the value y at *sym* and for all other arguments has the same value as f. If x is a sequence and y is some element then $x \circ y$ is the sequence obtained by appending y to the end of x. If x is a sequence then $cdr(x)$ is the sequence obtained from x by deleting its first element.

2.2. Syntactic Entities. We define the following syntactic entities: *Vars*, *Consts*, *Exprs*, *Asgns*, *SendExprs*, *RecExprs* and *RunExprs*.

Vars is the set of all variable names.

Consts is the set of all constants.

The term *var* refers to an element of *Vars*, *const* refers an element of *Consts* and *convar* refers to an element of either *Vars* or *Consts*.

Exprs is the set of expressions.

The term *expr* refers to an element of *Exprs*. The elements of *Exprs* are defined by the context-free grammar:

$bop ::= + \mid - \mid * \mid > \mid < \mid EQ \mid \mathrm{NE} \mid \mathrm{AND} \mid \mathrm{OR}$

$uop ::= - \mid !$

$expr ::= convar \mid (expr) \mid expr\; bop\; expr \mid uop\; expr \mid$
$\mathrm{LEN}\; var \mid timeout \mid var?[convar]$

The next four sets define different types of statements, or *state transformers*.

Asgns is the set of assignment statements, with elements of the form '$var = expr$' .

SendExprs is the set of send statements, with elements of the form '$var!expr$' .

RecExprs is the set of receive statements, with elements of the form '$var?convar$' .

RunExprs is the set of run statements, with elements of the form *run Name expr*' .

DEFINITION 2.1. *A* **primitive statement** *is an element of one of the sets: Exprs, Asgns, SendExprs, RecExprs, or RunExprs.*

Note that we have restricted *run* statements to a single parameter (in addition to the *proctype* name) and similarly we have restricted send and receive operations (and thereby channel declarations) to a single message field. These restrictions are meant to improve the readability of the definitions that follow.

2.3. Data Types and Value Domains. *BTypes* is the set of *basic* types, which we restrict here to $\{bit, byte, short, int\}$. *Types* is the set of types given by $BTypes \cup \{chan\}$.

To ensure that only finite state programs are defined, PROMELA sets finite upper bounds on certain parameters. We introduce symbolic names for these upper bounds. *MaxChanCap* is the maximum number of messages a channel can hold (all channels have finite capacity). *MaxNumProc* is maximum number of processes that can be active at a time, and *MaxNumChan* is the maximum number of channels that can be accessible at a time. For notational convience, we also define the following sets.

$MaxChanCapSet = \{\, i \mid 0 \le i < MaxChanCap \wedge i \in Naturals \,\}$

$MaxNumChanSet = \{\, i \mid 0 \le i \le MaxNumChan \wedge i \in Naturals \,\}$

$MaxNumProcSet = \{\, i \mid 0 \le i < MaxNumProc \wedge i \in Naturals \,\}$

The function *ValDomMap* specifies the values a variable of a given type can take during the execution of a program (Page 93 of [**Hol91**]).

ValDomMap is a map from *BTypes* to $2^{Integers}$ defined as follows:

$ValDomMap(bit) = \{\, i \mid 0 \le i \le 1 \wedge i \in Naturals \,\}$

$ValDomMap(byte) = \{\, i \mid 0 \le i \le 255 \wedge i \in Naturals \,\}$

$ValDomMap(short) = \{\, i \mid -2^{15} \le i \le 2^{15} - 1 \wedge i \in Integers \,\}$

$ValDomMap(int) = \{\, i \mid -2^{31} \le i \le 2^{31} - 1 \wedge i \in Integers \,\}$

$ValDomMap(chan) = MaxNumChanSet$

The above definition states, for example, that a variable of type *bit*, can take only one of the two values 0 or 1.

2.4. Labeled Transition Systems. A *symbolic labeled transition system* is the mathematical representation of the information contained in the body of a process type declaration (i.e., a PROMELA *proctype*).

DEFINITION 2.2. *A* **symbolic labeled transition system** *is a tuple of the form*

$$\langle Name, Structure, Start, locals, ChansOwned, Active, Param\rangle$$

where

- *Name is an element of Vars* (the name of the proctype).
- *Structure is* $\langle lstates, Act, \longmapsto\rangle$ *where lstates is a finite set of local states and transition relation* $\longmapsto \subseteq lstates \times Act \times lstates$. *Act is* $PLabels \times EModes \times Priorities$ *and its elements will be ranged over by* $\alpha, \beta, \cdots$. *PLabels is the set of primitive statements, i.e., the union of Exprs, Asgns, SendExprs, RecExprs and RunExprs. EModes is the set of execution modes* $\{normal, atomic, deterministic\}$. *Priorities is the set of natural numbers.*
  Informally, *Structure* encodes the control-flow of the sequential program named *Name* as a labeled directed graph. The nodes of this graph are the elements of set *lstates*. Each node represents a distinct local state of *Name*. Relation $\longmapsto$ represents the labeled edges between the nodes. If $\langle n_1, \alpha, n_2\rangle \in \longmapsto$ then it is possible to move to state n_2 from state n_1 by executing the primitive statement $\alpha.1$. The label of the edge also contains information about whether the statement $\alpha.1$ is to executed *normally*, *atomically* or *deterministically*. We call this the execution mode, and encode it as $\alpha.2$. Finally, $\alpha.3$ represents the *priority* of the transition. Priorities will be used to properly encode *unless* statements. A transition with a lower priority will not be selected for execution if one with strictly higher priority is also executable.
- *Start is an element of lstates.*
  It represents the initial state of the proctype.
- *locals is a* partial *function from the Vars to* $\cup\{\ ValDomMap(\tau) \mid \tau \in BTypes\}$.
  The partial function *locals* identifies the local variables of *Name*. Note that the function *locals* is partial; given a variable name *var*, *locals*(*var*) is undefined if *var* is not a local variable of *Name*. If *var* is indeed a local variable, then *locals*(*var*) is 0 (the default value, it it is uninitialized) or the value to which it is initialized. *locals* does not contain information about channel variables.
- *ChansOwned is a finite* sequence *of elements from*
  $$(Vars \times (MaxChanCapSet \cup \{\Omega\}) \times (Types \cup \{\Omega\}).$$
  The entity *ChansOwned* records the information about the channel variables accessed by this sequential program; specifically, it records the capacity of each channel (the maximum number of messages the individual channel can hold) and the data type of its message field. For an uninitialized variable of type *chan* these values above are set to Ω.
- *Active is an integer which is either 0 or 1.*
- *Param is a member of Vars.*

This is a single formal parameter that is bound to a value when the process is instantiated.

(End of definition 2.2.)

2.5. Local and Global States. *Lstates* is the set of *local states* (i.e., process states). *Gstates* is the set of *global states* (i.e., system states). Formally, a *global state* Σ is a pair of the form $\langle X, g\rangle$. X is a finite set of elements of the form $\langle \mathit{Name}, \mathit{pid}, \mathit{pc}, \mathit{locals}, \mathit{lChansOwned}\rangle$. Each process active in the state Σ is represented by some member of X exactly once.

1. *Name* is an element of *Vars*.
   It is the name of the proctype corresponding to this process. Note that there could be more than one active process corresponding to the same proctype.
2. *pid* belongs to $MaxNumProcSet$.
   It is the unique identifier corresponding to an active process. In other words, if x and x' are two *distinct* members of X then $x.2 \neq x'.2$. Thus any member of X can be uniquely identified by *pid*. Thus we denote $X : i$ to denote that particular element of x of X such that $x.\mathit{pid} = i$.
3. *pc* belongs to *Lstates*
   It is the program counter of the active process; it points to some local state.
4. *locals* is a partial function from *Vars* to $\cup\{\ \mathit{ValDomMap}(\tau) \mid \tau \in \mathit{BTypes}\}$.
   It is the set of local variables of the active process.
5. *lChansOwned* is a finite sequence of elements from
   $(\mathit{Vars} \times (\mathit{MaxChanCapSet} \cup \{\Omega\}) \times (\mathit{Types} \cup \{\Omega\})$.
   It is the list of channels owned by this active process.

The second of component g of the global state Σ is a tuple of the kind $\langle \mathit{globals}, \mathit{gChansOwned}, \mathit{NrProcs}, \mathit{Handshake}, \mathit{Exclusive}, \mathit{ChIdMap}, \mathit{ChContMap}, \mathit{timeout}, \mathit{dstep}\rangle$ where

1. *globals* is a partial function from *Vars* to $\cup\{\ \mathit{ValDomMap}(\tau) \mid \tau \in \mathit{BTypes}\}$
   It is the set of global variables.
2. *gChansOwned* is a finite sequence of elements from
   $(\mathit{Vars} \times (\mathit{MaxChanCapSet} \cup \{\Omega\}) \times (\mathit{Types} \cup \{\Omega\})$.
   It is the list of globally declared channels.
3. *NrProcs* is an element of *MaxNumProcSet.*
   $\mathit{NrProcs} + 1$ is the number of processes currently active at this global state. It cannot exceed *MaxNumProc.*
4. *Handshake* is either an empty set or an element of $\mathit{MaxNumProcSet} \times \mathit{MaxNumChanSet} \times \mathit{Consts}$.
   If *Handshake* is not empty it means that there is a synchronous communication (a rendezvous handshake) in progress; the active process with process-id *Handshake*.1 is sending the value *Handshake*.3 over the channel *Handshake*.2.
5. *Exclusive* is either empty or an element of *MaxNumProcSet.*
   If *Exclusive* is non-empty the process with process-id *Exclusive* is executing atomically or deterministically, and cannot be interrupted by other process executions.

6. *ChIdMap* is a partial function from $(MaxNumProcSet \cup \{-1\}) \times Vars$ to *MaxNumChanSet*.
   This partial function assign an unique channel-id to every every active channel; given a process-id and a variable name as arguments this function provides the channel-id. Global channels are assigned process-id -1. If the variable name is not a channel name for that process-id then the function is undefined on these arguments.
7. *ChContMap* is a partial function from *MaxNumChanSet* to

   $$\bigcup\{\ [\cup\{\ ValDomMap(\tau) \mid \tau \in Types\}]^i \mid 0 \leq i \leq MaxChanCap\}$$

   Given a channel-id, this maps provides the current contents of that channel.
8. *timeout* and *dstep* are integers which take value either 0 or 1.
   If all processes in the system are blocked (i.e., there is no executable statement) when the value of *timeout* equals 0, then the value of changes to 1 for one execution step, and returns to its default value 0. If *dstep* equals 1, some process is executing deterministically.

2.6. Auxiliary Functions. This subsection is devoted a list of auxiliary functions that are used for querying and updating global states.

2.6.1. *Characteristic Functions.* When the same identifier is used to define both a global variable and a local variable, the usual scoping rules determine which variable is intended. The following functions can be used to resolve these issues.

Islchan is a function defined from $Vars \times MaxNumProcSet \times Gstates$ to the set of truth values { *tt*,*ff* } as follows. $Islchan(var, pid, \langle X, g\rangle)$ is *tt* if there exists an x in $(X : pid).lChansOwned$ such that $x.1$ equals *var* and *ff* otherwise. Intuitively, $Islchan(var, pid, \langle X, g\rangle)$ is true iff in the global state $\langle X, g\rangle$, *var* is a local channel owned by a process whose process-id is *pid*.

Islnumvar is a function defined from $Vars \times MaxNumProcSet \times Gstates$ to the set of truth values { *tt*,*ff* } as follows. $Islnumvar(var, pid, \langle X, g\rangle)$ is *tt* if $(X : pid).locals(var)$ is well defined and *ff* otherwise.

Isgchan is a partial function defined from $Vars \times MaxNumProcSet \times Gstates$ to the set of truth values { *tt*,*ff* } as follows. $Isgchan(var, pid, \langle X, g\rangle)$ is *tt* if $Islnumvar(var, pid, (X : pid).)$ as well $Islchan(var, pid, \langle X, g\rangle)$ are *ff* and there exists an x in $g.gChansOwned$ such that $x.1$ equals *var* and *ff* otherwise. Intuitively, a variable is a global channel variable only if there is no local numeric or local channel variable of the same name.

Ischan is a function defined from $Vars \times MaxNumProcSet \times Gstates$ to the set of truth values { *tt*,*ff* } as follows. $Ischan(var, pid, \Sigma)$ is *tt* if either $Isgchan(var, pid, \Sigma)$ is *tt* or $Islchan(var, pid, \Sigma)$ is *tt* and *ff* otherwise.

Isgnumvar is a function defined from $Vars \times MaxNumProcSet \times Gstates$ to the set of truth values { *tt*,*ff* } as follows. $Isgnumvar(var, pid, \langle X, g\rangle)$ is *tt* if $g.globals(var)$ is well defined and $Islnumvar(var)$ is *ff* and $Islchan(var)$ is *ff*. Otherwise it is *ff*.

Isnumvar is a partial function defined from $Vars \times MaxNumProcSet \times Gstates$ to the set of truth values { *tt*,*ff* } as follows. $Isnumvar(var, pid, \Sigma)$ is *tt* if either $Isgnumvar(var, pid, \Sigma)$ is *tt* or $Islnumvar(var, pid, \Sigma)$ is *tt* and *ff* otherwise.

2.6.2. *Other Auxiliary Functions.* We define some more auxiliary functions that depend on the characteristic functions. The function *ChanId*, given a variable name and a process-id provides the unique channel-id associated with it if it is indeed a channel. Its definition makes use of the characteristic functions to ensure the proper scoping rule is enforced. Similary the function *RightVal*, given a variable and a process-id provides the right value associated with the variable; if it is a channel variable then its channel-id is returned.

ChanId is a partial function defined from $Vars \times MaxNumProcSet \times Gstates$ to the set $(MaxNumChanSet \cup \Omega)$ as follows. $ChanId(var, pid, \Sigma)$ is Ω if $Ischan(var, pid, \Sigma)$ is *ff*. If $Islchan(var, pid, \Sigma)$ is *tt* then $ChanId(var, pid, \Sigma)$ is $ChIdMap(\langle pid, var\rangle)$. If $Isgchan(var, pid, \Sigma)$ is *tt* then $ChanId(var, pid, \Sigma)$ is $ChIdMap(\langle -1, var\rangle)$.

RightVal is a partial function defined from $Vars \times MaxNumProcSet \times Gstates$ to the set *Integers* as follows.

- If $Islchan(var, pid, \Sigma)$ is *tt* then $RightVal(var, pid, \Sigma)$ is $ChanId(var, pid, \Sigma)$.
- If $Islnumvar(var, pid, \Sigma)$ is *tt* then $RightVal(var, pid, \Sigma)$ is $locals(var)$.
- If $Isgnumvar(var, pid, \Sigma)$ is *tt* then $RightVal(var, pid, \Sigma)$ is $globals(var)$.
- If $Isgchan(var, pid, \Sigma)$ is *tt* then $RightVal(var, pid, \Sigma)$ is $ChanId(var, -1, \Sigma)$.

capacity is a partial function defined from $Vars \times MaxNumProcSet \times Gstates$ to the set *MaxChanCapSet* as follows. If $Islchan(var, pid, \Sigma)$ is *tt* then $capacity(var, pid, \Sigma)$ is $x.2$ where x is a element of the sequence *lChansOwned* such that $x.1$ is same as *var*. If $Isgchan(var, pid, \Sigma)$ is *tt* then $capacity(var, pid, \Sigma)$ is $x.2$ where x is a element of the sequence *gChansOwned* such that $x.1$ is same as *var*.

$compatible(var_1, var_2, pid, \Sigma)$ is *tt* if $RightVal(var_1, pid, \Sigma)$ is same as $RightVal(var_2, pid, \Sigma)$ and *ff* otherwise. $compatible(const, var, pid, \Sigma)$ is *tt* if $RightVal(var_1, pid, \Sigma) = const$ and *ff* otherwise. $compatible(const_1, const_2)$ is *tt* if $const_1 = const_2$. Synchronous communication can happen only if there is some compatibility between the sender and the receiver; the above function formalizes this relationship.

2.6.3. *Evaluation of Expressions.* We define the function $[\![\ ]\!]_{pid,\Sigma}$ to provides the value of the given expression. This function returns Ω if an error occurs during the evaluation.

The function $[\![\ ]\!]_{pid,\Sigma}$ from *Exprs* to $Integers \cup \{\Omega\}$ is defined as follows. Let $\Sigma = \langle X, g\rangle$. If there is no element x of X with $x.2$ equals *pid* then $[\![\ ]\!]_{pid,\Sigma}$ is a constant function with value Ω. Otherwise let x be the element $\langle Name, pid, pc, locals, lChansOwned\rangle$. Now $[\![expr]\!]_{pid,\Sigma}$ is given by structural induction below. For readability, we omit the subscripts for $[\![\ ]\!]$ below.

- If $expr \equiv const$ then $[\![expr]\!] = const$.
- If $expr \equiv timeout$ then $[\![expr]\!] = g.timeout$.
- If $expr \equiv var$ then $[\![expr]\!] = RightVal(var, pid, \Sigma)$.
- If $expr = \mathrm{LEN}\, var$ and $Ischan(var, pid, \Sigma) = ff$ then $[\![expr]\!] = \Omega$.
- If $expr = \mathrm{LEN}\, var$ and $Ischan(var, pid, \Sigma) = tt$ and $[\![var]\!] = 0$ then $[\![expr]\!] = \Omega$.
- If $expr = \mathrm{LEN}\, var$ and $Ischan(var, pid, \Sigma) = tt$ and $[\![var]\!] \neq 0$ and $ChContMap([\![var]\!])$ is undefined then $[\![expr]\!] = \Omega$

- If $expr$ = LENvar and $Ischan(var, pid, \Sigma)$ = tt and $[\![var]\!] \neq 0$ and $ChContMap([\![var]\!])$ is well defined then $[\![expr]\!] = |ChContMap([\![var]\!])|$.
- If $expr = var?[convar]$ and ($Ischan(var) = ff$ or $[\![\mathrm{LEN}var]\!] = \Omega$) then $[\![expr]\!] = \Omega$
- If ($expr = var?[convar]$ and $[\![\mathrm{LEN}var]\!] = 0$ then $[\![expr]\!] = 0$
- If $expr = var?[convar]$ and $Ischan(var) = tt$ and $[\![\mathrm{LEN}var]\!] \notin \{0, \Omega\}$ and $compatible(first(var_1, pid, \Sigma), [\![convar]\!], pid, \Sigma) = tt$ then $[\![expr]\!] = 1$
- If $expr = var?[convar]$ and $Ischan(var) = tt$ and $[\![\mathrm{LEN}var]\!] \notin \{0, \Omega\}$ and $compatible(first(var_1, pid, \Sigma), [\![convar]\!], pid, \Sigma) = ff$ then $[\![expr]\!] = 0$
- If $expr = (expr_1)$ then $[\![expr]\!] = [\![expr_1]\!]$
- If $expr$ = uop $expr_1$ then $[\![expr]\!] = [\![uop]\!]([\![expr_1]\!])$.
  If $expr = expr_1$ bop $expr_2$ then $[\![expr]\!] = [\![bop]\!]([\![expr_1]\!], [\![expr_2]\!])$.

first is a partial function defined from $Vars \times MaxNumProcSet \times Gstates$ to the set *Integers* as follows. $first(var, pid, \Sigma) = ChContMap([\![var]\!]).1.2$.

2.6.4. *Execution Errors, Failure.* As an example of an execution error, consider the statement $x = 5$, where x is channel variable. The function *WillFail* formally describes these situations.

WillFail is a partial function defined from $PLabels \times MaxNumProcSet \times Gstates$ to the set of truth values { *tt*,*ff* } as follows.

- $WillFail(expr, pid, \Sigma)$ is *tt* if $[\![expr]\!]$ is Ω and *ff* otherwise.
- $WillFail(var = expr, pid, \Sigma)$ is *tt* if ($[\![expr]\!]$ is Ω) or (exactly one of $Ischan(var)$ and $Ischan(expr)$ is *tt*). Otherwise $WillFail(var = expr, pid, \Sigma)$ is *ff*.
- $WillFail(var!expr, pid, \Sigma)$ is *tt* if $[\![expr]\!]$ is Ω or ($Ischan(var)$ is *ff*) or ($Ischan(var)$ is $tt \wedge ChContMap(ChanId(var, pid, \Sigma))$ is undefined). Otherwise $WillFail(var!expr, pid, \Sigma)$ is *ff*.
- $WillFail(var?convar, pid, \Sigma)$ is *tt* if ($[\![convar]\!]$ is Ω) or ($Ischan(var)$ is *ff*) or ($Ischan(var)$ is *tt* and $ChanId(var, pid, \Sigma) = 0$) or ($Ischan(var)$ is $tt \wedge ChContMap(ChanId(var, pid, \Sigma))$ is undefined. Otherwise $WillFail(var?convar, pid, \Sigma)$ is *ff*.

2.6.5. *Channel Manipulation Functions.* When a new process is created, all the channels owned by it are entered in the *ChIdMap*. Similary when a process terminates, it local channels are removed from the *ChIdMap*. The functions defined in this subsubsection accomplish this task.

DEFINITION 2.3. *Incorp(ChIdMap, ChansOwned, pid) is defined as a partial function f from the set $(MaxNumProcSet \cup \{-1\}) \times Vars$ to the set MaxNumChanSet. f has the same value on those elements at which ChIdMap is well defined. Further if $\langle var, \Omega, \Omega\rangle$ is a member of ChansOwned then $f(\langle pid, var\rangle)$ is defined as 0. Let $c_1 \cdots c_k$ be the sequence ChansOwned with no elments of the form $\langle var, \Omega, \Omega\rangle$. Let m be the maximum value taken by ChIdMap. Then $f(pid, c_i.1)$ is defined to be $m + i$ for $1 \leq i \leq k$.*

The function *Incorp* augments the information in the *ChIdMap* with that present in *ChansOwned* to produce another function that is of the same type as *ChIdMap*.

DEFINITION 2.4. *IncorpCont(ChContMap, ChIdMap, ChansOwned, pid) is a partial function f that has the same domain and codomain as that* *of ChContMap. $f(x)$ is defined as an empty sequence if there exists a $c \in$ ChansOwned such that ChIdMap(pid, c) $= x$ and $x \neq 0$. At other arguments f and ChContMap are identical.*

The following two functions remove information from the *ChIdMap* and *ChContMap*.

DEFINITION 2.5. *RetractId(ChIdMap, ChansOwned, pid) is a partial function f from the set (MaxNumProcSet $\cup \{-1\}) \times$ Vars to MaxNumChanSet. $f(pid, c)$ is undefined if there exists a $x \in$ ChansOwned such that $x.1 = c$. At other arguments, f and ChIdMap are identical.*

DEFINITION 2.6. *RetractCont(ChContMap, ChIdMap, ChansOwned, pid) is a partial function f that has the same domain and codomain as that of ChContMap. $f(x)$ is undefined if there exists a $c \in$ ChansOwned such that ChIdMap(pid, c) $= x$. At other arguments f and ChContMap are identical.*

3. Preconditions

We refer to the directed labeled edges of a symbolic transition system as transitions. This section gives the definition of a *transition* and a *current transition.* A current transition is *executable* in a global state only if certain conditions are satisfied. This section formalizes these notions.

DEFINITION 3.1. *If S is a symbolic labeled transition system with $S.1 =$ Name and $S.2.3$ contains $\langle s, \alpha, t\rangle$ then $\langle Name, s, \alpha, t\rangle$ will be called a transition.*

Informally, a transition is *current* in a global state Σ if the program counter of some active sequential program points to the source state of the transition.

DEFINITION 3.2. *A* current *transition of a state $\Sigma = \langle X, g\rangle$ is the tuple $\langle pid, Name, s, \alpha, t\rangle$ if the following hold:*

1. *$\langle Name, s, \alpha, t\rangle$ is a transition.*
2. *$\langle Name, pid, s, locals, lChansOwned\rangle$ is a member of X for some locals and lChansOwned.*

Next, we define when a current transition is said to be *executable.*

DEFINITION 3.3. *A current transition $\langle Name, pid, s, \alpha, t\rangle$ is* executable *in the state $\Sigma = \langle X, g\rangle$ if all of the following 10 conditions hold.*
1. $g.Exclusive = \{\}$ or $g.Exclusive = \{pid\}$.
Intuitively, this condition states that in order for a current transition to be executable no *other* process may have exclusive control.
2. $WillFail(\alpha.label, pid, \Sigma) = tt$
$\Longrightarrow g.Handshake = \{\}$.
If a synchronous communication is in progress, a current transition is not permitted to execute and cause an error.
3. $(\alpha.label \equiv expr) \wedge WillFail(\alpha.label, pid, \Sigma) = ff$
$\Longrightarrow$
$g.Handshake = \{\} \wedge ([\![expr]\!]_{\Sigma,pid} \neq 0)$.
An expression is executable if it value is nonzero and if no synchronous communication is in progress.

4. $\alpha.label \equiv var!expr \wedge WillFail(\alpha.label, pid, \Sigma) = ff \wedge capacity(var, pid, \Sigma) \neq 0$
$\Longrightarrow$
$g.Handshake = \{\} \wedge [\![LENvar]\!]_{pid,\Sigma} < capacity(var, pid, \Sigma)$.
The statement that sends a value of the *expr* to the channel *var* is executable only if the the channel *var* is not full. (see Page 99 of [**Hol91**])
5. $\alpha.label \equiv var?convar \wedge WillFail(\alpha.label, pid, \Sigma) = ff$
$\wedge\ capacity(var, pid, \Sigma) \neq 0$
$\Longrightarrow$
$g.Handshake = \{\} \wedge [\![var?[convar]]\!]_{pid,\Sigma} = 1$.
Similarly the statement that extracts a value from the channel *var* is executable only if the channel *var* is nonempty (see Page 100 of [**Hol91**]).
6. $\alpha.label \equiv var!expr \wedge WillFail(\alpha.label, pid, \Sigma) = ff \wedge capacity(var, pid, \Sigma) = 0$
$\Longrightarrow$
$g.Handshake = \{\} \wedge$
$\exists qid, Name', s', t', x, \beta :\ qid \neq pid \wedge \beta.label \equiv var'_1?convar \wedge$
$[\![var'_1]\!]_{\Sigma,qid} = [\![var]\!]_{\Sigma,pid} \wedge \langle qid, Name', s', \beta, t'\rangle$ *is a current transition of* $\Sigma \wedge$
$compatible(convar, [\![expr]\!]_{g,pid}, pid, \Sigma)$.
A synchronous-send statement on a channel *var* is executable only if there is another distinct process that has synchronous-receive statement on the same channel as a current transition. (see Page 100 of [**Hol91**])
7. $\alpha.label \equiv var?convar \wedge WillFail(\alpha.label, pid, \Sigma) = ff \wedge$
$capacity(var, pid, \Sigma) = 0$
$\Longrightarrow$
$\exists qid, const :\ qid \neq pid \wedge g.Handshake = \{\langle qid, [\![var]\!]_{pid,\Sigma}, const\rangle\}$
$\wedge\ compatible(convar, const, pid, \Sigma)$.
8. $(\alpha.label \equiv asgn) \wedge WillFail(\alpha.label, pid, \Sigma) = ff$
$\Longrightarrow$
$g.Handshake = \{\}$.
9. $(\alpha.label \equiv run\ Name\ arg) \wedge WillFail(\alpha.label, pid, \Sigma) = ff$
$\Longrightarrow$
$g.Handshake = \{\} \wedge g.NrProcs < MaxNumProc - 1$.
A new process can be created only if the number of currently active processes does not exceed the preset upper bound.
10. $(\alpha.label \equiv end)$
$\Longrightarrow$
$g.Handshake = \{\} \wedge pid = g.NrProcs$.
Processes can only terminate in the reverse order of their creation (i.e., a process cannot terminate until all its children have terminated first). If processes are issued pid's sequentially starting from zero, then the only process that can terminate is the one that with a pid equal to *g.NrProcs*. (see Page 96 of [**Hol91**])

4. Effects

In this section we define how each of the primitive statements transforms the current global state when it is executed.

4.1. Normal Execution Mode. The effect of the execution of a statement depends in part on the "mode" in which it is executed. The following definition presents the effects of statements executed in "normal" mode (i.e., not involving *atomic* or *dstep* sequences).

DEFINITION 4.1. *If* $\langle pid, Name, s, \alpha, t\rangle$ *is* executable *in the state* $\Sigma = \langle X, g\rangle$ *with* $\alpha.mode = normal$ *then* $Effect(\Sigma, \langle pid, Name, s, \alpha, t\rangle)$ *is* $\Sigma' = \langle X', g'\rangle$ *where*
1. $WillFail(\alpha.label, pid, \Sigma) = tt \implies \Sigma' = \Omega$
We use the symbol Ω to denote the global error state. The above condition states that the execution of an error-producing statement will transform any state into the error state.
2. $(\alpha.label \equiv expr) \wedge WillFail(\alpha.label, pid, \Sigma) = ff \wedge [\![expr]\!]_{pid,\Sigma} \neq 0$
$\implies$
$X' = X[\langle Name, pid, t, locals, lChansOwned\rangle /$
$\langle Name, pid, s, locals, lChansOwned\rangle]$
$g' = g[Exclusive = \{\}, dstep = 0]$
The program counter of the process is updated. No exclusive execution control is obtained or preserved by any statement that is executed in normal mode.
3. $(\alpha.label \equiv var = expr \ \wedge \ WillFail(\alpha.label, pid, \Sigma) = ff)$
$\implies$
$X' = X[\langle Name, pid, t, locals, lChansOwned\rangle /$
$\langle Name, pid, s, locals', lChansOwned\rangle]$
$g' = g[Exclusive = \{\}, dstep = 0, globals'/globals,$
$ChIdMap'/ChIdMap]$ *where*

- $islnumvar(var) = tt$
  $\implies (locals' = locals[var = [\![expr]\!]_{pid,\Sigma}], globals' = globals,$
  $ChIdMap' = ChIdMap)$
- $isgnumvar(var) = tt$
  $\implies (globals' = globals[var = [\![expr]\!]_{pid,\Sigma}], locals' = locals,$
  $ChIdMap' = ChIdMap)$
- $ischan(var) = tt$
  $\implies (locals' = locals, globals' = globals,$
  $ChIdMap' = ChIdMap[ChanId(var, pid, \Sigma) = [\![expr]\!]_{pid,\Sigma}])$

The effect of this statement is to assign the value of the expression to the variable. If the variable is a channel variable then the *ChIdMap* is updated. The scope rules determine if the variable is local or global.
4. $(\alpha.label \equiv var!expr \wedge capacity(var, pid, \Sigma) \neq 0 \wedge WillFail(\alpha.label, pid, \Sigma) = ff)$
$\implies$
$X' = X[\langle Name, pid, t, locals, lChansOwned\rangle /$
$\langle Name, pid, s, locals, lChansOwned\rangle]$
$g' = g[Exclusive = \{\}, dstep = 0, ChContMap'/ChContMap]$ *where*
$ChContMap'$ *is given by* $ChContMap[ChIdMap(pid, var, \Sigma) =$
$ChContMap(ChIdMap(pid, var, \Sigma)) \circ [\![expr]\!]_{pid,\Sigma}]$
To understand the above condition, recall the meaning of the notation $f[x = y]$ from Subsection 2.1; the value of the expression is appended at the end of the contents of the channel *var*. (See Page 98 of [**Hol91**]).

5. $(\alpha.label \equiv var?const \wedge capacity(var, pid, \Sigma) \neq 0$
$\wedge\ WillFail(\alpha.label, pid, \Sigma) = ff)$
$\Longrightarrow$
$X' = X[\langle Name, pid, t, locals, lChansOwned\rangle /$
$\langle Name, pid, s, locals, lChansOwned\rangle]$
$g' = g[Exclusive = \{\}, dstep = 0, ChContMap'/ChContMap]$ *where*
ChContMap′ is given by $ChContMap[ChIdMap(pid, var, \Sigma) =$
$cdr(ChContMap(ChIdMap(pid, var, \Sigma)))]$
Since this transition is executable it is necessarily the case that the head of the channel *var* is *const*.
6. $(\alpha.label \equiv var_1?var_2 \wedge capacity(var_1, pid, \Sigma) \neq 0$
$\wedge\ WillFail(\alpha.label, pid, \Sigma) = ff)$
$\Longrightarrow$
$X' = X[\langle Name, pid, t, locals, lChansOwned\rangle /$
$\langle Name, pid, s, locals', lChansOwned\rangle]$
$g' = g[Exclusive = \{\}, globals'/globals, ChIdMap'/ChIdMap,$
$dstep = 0, ChContMap'/ChContMap]$ *where*

- $islnumvar(var_1) = tt$
  $\Longrightarrow (locals' = locals[var_2 = first(var_1, pid, \Sigma)],$
  $globals' = globals, ChIdMap' = ChIdMap)$
- $ischan(var) = tt$
  $\Longrightarrow (locals' = locals, globals' = globals,$
  $ChIdMap' = ChIdMap[ChanId(var_2, pid, \Sigma) = first(var_1, pid, \Sigma)])$
- $isgnumvar(var) = tt$
  $\Longrightarrow (globals' = globals[var_2 = first(var_1, pid, \Sigma)], locals' = locals,$
  $ChIdMap' = ChIdMap)$
- $ChContMap' =$
  $ChContMap[ChIdMap(pid, var, \Sigma) =$
  $cdr(ChContMap(ChIdMap(pid, var, \Sigma)))]$

The value at the head of the channel var_1 is removed and assigned to the value var_2, just like in a normal PROMELA assignment statement.
7. $(\alpha.label \equiv var!expr \wedge capacity(var, pid, \Sigma) = 0 \wedge WillFail(\alpha.label, pid, \Sigma) =$
$ff)$
$\Longrightarrow$
$X' = X[\langle Name, pid, t, locals, lChansOwned\rangle /$
$\langle Name, pid, s, locals, lChansOwned\rangle]$
$g' = g[Exclusive = \{\}, dstep = 0,$
$Handshake = \{\langle pid, ChanId(var, pid, \Sigma), [\![expr]\!]_{pid,\Sigma}\rangle\}]$
The effect of a synchronous-send statement is to store the following values in the mathematical structure *Handshake*: the pid of the sender process, the channel-id of the synchronous channel and the value of the expression being sent. This signifies the initiation of a synchronous (rendezvous hand-shake) communication.
8. $(\alpha.label \equiv var?const \wedge capacity(var, pid, \Sigma) = 0 \wedge WillFail(\alpha.label, pid, \Sigma) =$
$ff \wedge Handshake = \{\langle qid, [\![var]\!]_{pid,\Sigma}, const'\rangle\})$
$\Longrightarrow$
$X' = X[\langle Name, pid, t, locals, lChansOwned\rangle /$

$\langle Name, pid, s, locals, lChansOwned\rangle]$
$g' = g[Exclusive = \{\}, dstep = 0, Handshake = \{\}]$
Since this transition is executable, it is necessarily the case that $qid \neq pid$ and $const' = const$. One of the effect of this transition is to set structure *Handshake* to $\{\}$, which signifies the completion of the synchronous communication.
9. $(\alpha.label \equiv var_1?var_2 \wedge capacity(var_1, pid, \Sigma) = 0$
$\wedge\ WillFail(\alpha.label, pid, \Sigma) = ff)$
$\wedge\ Handshake = \{\langle qid, [\![var_1]\!]_{pid,\Sigma}, const\rangle\}$
$\Longrightarrow$
$X' = X[\langle Name, pid, t, locals, lChansOwned\rangle /$
$\langle Name, pid, s, locals', lChansOwned\rangle]$
$g' = g[Exclusive = \{\}, globals'/globals, dstep = 0, ChIdMap'/ChIdMap]$
where

- $islnumvar(var_2) = tt$
$\Longrightarrow (locals' = locals[var_2 = const], globals' = globals, ChIdMap' = ChIdMap)$
- $ischan(var_2) = tt$
$\Longrightarrow (locals' = locals, globals' = globals,$
$ChIdMap' = ChIdMap[ChanId(var_2, pid, \Sigma) = const])$
- $isgnumvar(var_2) = tt$
$\Longrightarrow (globals' = globals[var_2 = const], locals' = locals, ChIdMap' = ChIdMap)$

The effect of this statement is to assign the value in *Handshake* to the variable var_2 and to make the *Handshake* equal to $\{\}$ signifying the termination of the synchronous communication.
10. $(\alpha.label \equiv run\ Name_1\ arg) \wedge WillFail(\alpha.label, pid, \Sigma) = ff) \wedge$
$\langle Name_1, Structure_1, Start_1, locals_1, ChansOwned_1, Active_1, Param_1\rangle$
is a symbolic transition system $\Longrightarrow$
$X" = X[\langle Name, pid, t, locals, lChansOwned\rangle /$
$\langle Name, pid, s, locals, lChansOwned\rangle]$
$X' = X" \cup \langle Name_1, NrProcs + 1, Start_1, locals_2, ChansOwned_1\rangle$
$locals_2 = locals_1[Param_1 = [\![arg]\!]_{pid,\Sigma}]$
$g' = g[Exclusive = \{\}, NrProcs = NrProcs + 1, dstep = 0,$
$Incorp(ChIdMap, ChansOwned_1, NrProcs + 1)/ChIdMap,$
$IncorpCont(ChContMap, ChIdMap, lChansOwned_1, NrProcs + 1)/$
$ChContMap]$
The effect of this statement is to create a new process with proctype *Name*. The smallest unassigned nonnegative integer is assigned as its process-id; this value equals the old value of *NrProcs*, from before the statement is executed. The formal parameter is instantiated with the value of the actual parameter. *ChIdMap* and *ChContMap* are updated accordingly.
11. $(\alpha.label \equiv end \wedge WillFail(\alpha.label, pid, \Sigma) = ff)$
$\Longrightarrow$
$X' = X \setminus \{\langle Name, pid, t, locals, lChansOwned\rangle\}$
$g' = g[RetractId(ChIdMap, lChansOwned, pid)/ChIdMap,$
$RetractCont(ChContMap, ChIdMap, lChansOwned, pid)/ChContMap,$
$dstep = 0, NrProcs = NrProcs - 1]$

4.2. Atomic and Deterministic Execution Mode. Generally, when a process executes a primitive statement in *atomic* or *deterministic* mode it retains the exclusive right to execute also the next statement (it "retains control"). Otherwise, the effect of the execution of statements in these two modes is same as that in normal mode. To indicate this, in the following definition we set $Exclusive = \{pid\}$ besides the other conditions that are presented in the Definition 4.1. There is only one statement that when executed in *atomic* mode does *not* retain control: the synchronous send statement. Furthermore, it is an error to execute synchronous-send and synchronous-receive in *deterministic* mode.

The parts of the following definition that differ from the corresponding parts in Definition 4.1 are indicated in bold.

DEFINITION 4.2. *If* $\langle pid, Name, s, \alpha, t\rangle$ *is* executable *in the state* $\Sigma = \langle X, g\rangle$ *with* $\alpha.mode \in \{atomic, deterministic\}$ *then*
$Effect(\Sigma, \langle pid, Name, s, \alpha, t\rangle)$ *is* $\Sigma' = \langle X', g'\rangle$ *where*
1. $WillFail(\alpha.label, pid, \Sigma) = tt \implies \Sigma' = \Omega.$
2. $(\alpha.label \equiv expr) \wedge WillFail(\alpha.label, pid, \Sigma) = ff \wedge [\![expr]\!]_{pid,\Sigma} \neq 0$
$\implies$
$X' = X[\langle Name, pid, t, locals, lChansOwned\rangle /$
$\langle Name, pid, s, locals, lChansOwned\rangle]$
$g' = g[\mathbf{Exclusive} = \{pid\}, \mathbf{dstep} = dstep']$
$\mathbf{dstep}' = 1$ *if* $\alpha.label = deterministic$ *and 0 otherwise* .
3. $(\alpha.label \equiv var = expr \wedge WillFail(\alpha.label, pid, \Sigma) = ff)$
$\implies$
$X' = X[\langle Name, pid, t, locals, lChansOwned\rangle /$
$\langle Name, pid, s, locals', lChansOwned\rangle]$
$g' = g[\mathbf{Exclusive} = \{pid\}, globals'/globals, ChIdMap'/ChIdMap,$
$\mathbf{dstep}'/\mathbf{dstep}]$
where $dstep' = 1$ *if* $\alpha.label = deterministic$ *and* 0 *otherwise.*

- $islnumvar(var) = tt$
  $\implies (locals' = locals[var = [\![expr]\!]_{pid,\Sigma}], globals' = globals,$
  $ChIdMap' = ChIdMap)$
- $ischan(var) = tt$
  $\implies (locals' = locals, globals' = globals, ChIdMap'$
  $= ChIdMap[ChanId(var, pid, \Sigma = [\![expr]\!]_{pid,\Sigma}])$
- $isgnumvar(var) = tt$
  $\implies (globals' = globals[var = [\![expr]\!]_{pid,\Sigma}], locals' = locals,$
  $ChIdMap' = ChIdMap)$

4. $(\alpha.label \equiv var!expr \wedge capacity(var, pid, \Sigma) \neq 0 \wedge WillFail(\alpha.label, pid, \Sigma) = ff)$
$\implies$
$X' = X[\langle Name, pid, t, locals, lChansOwned\rangle /$
$\langle Name, pid, s, locals, lChansOwned\rangle]$
$g' = g[\mathbf{Exclusive} = \{pid\}, ChContMap'/ChContMap, dstep/dstep']$
$dstep' = 1$ *if* $\alpha.label = deterministic$ *and* 0 *otherwise. where*
$ChContMap' = ChContMap[ChIdMap(pid, var, \Sigma) =$
$ChContMap(ChIdMap(pid, var, \Sigma)) \circ [\![expr]\!]_{pid,\Sigma}].$

5. $(\alpha.label \equiv var?const \wedge capacity(var, pid, \Sigma) \neq 0$
$\wedge\ WillFail(\alpha.label, pid, \Sigma) = ff)$
$\Longrightarrow$
$X' = X[\langle Name, pid, t, locals, lChansOwned\rangle /$
$\langle Name, pid, s, locals, lChansOwned\rangle]$
$g' = g[\mathbf{Exclusive} = \{pid\}, ChContMap'/ChContMap, \mathbf{dstep/dstep'}]$
where

- $\mathbf{dstep'} = 1$ *if* $\alpha.label = deterministic$ *and* 0 *otherwise.*
- $ChContMap' =$
  $ChContMap[ChIdMap(pid, var, \Sigma) =$
  $cdr(ChContMap(ChIdMap(pid, var, \Sigma)))]$.

6. $(\alpha.label \equiv var_1?var_2 \wedge capacity(var_1, pid, \Sigma) \neq 0$
$\wedge\ WillFail(\alpha.label, pid, \Sigma) = ff)$
$\Longrightarrow$
$X' = X[\langle Name, pid, t, locals, lChansOwned\rangle /$
$\langle Name, pid, s, locals', lChansOwned\rangle]$
$g' = g[\mathbf{Exclusive} = \{pid\}, globals'/globals, ChIdMap'/ChIdMap,$
$ChContMap'/ChContMap, \mathbf{dstep/dstep'}]$ *where*

- $\mathbf{dstep'} = 1$ *if* $\alpha.label = deterministic$ *and* 0 *otherwise.*
- $islnumvar(var_1) = tt$
  $\Longrightarrow (locals' = locals[var_2 = first(var_1, pid, \Sigma)],$
  $globals' = globals, ChIdMap' = ChIdMap)$
- $ischan(var) = tt$
  $\Longrightarrow (locals' = locals, globals' = globals,$
  $ChIdMap' = ChIdMap[ChanId(var_2, pid, \Sigma) = first(var_1, pid, \Sigma)]$
- $isgnumvar(var) = tt$
  $(globals' = globals[var_2 = first(var_1, pid, \Sigma)], locals' = locals,$
  $ChIdMap' = ChIdMap)$
- $ChContMap' = ChContMap[ChIdMap(pid, var, \Sigma) =$
  $cdr(ChContMap(ChIdMap(pid, var, \Sigma)))]$

7. $(\alpha.label \equiv var!expr \wedge capacity(var, pid, \Sigma) = 0$
$\wedge\ WillFail(\alpha.label, pid, \Sigma) = ff)$
$\Longrightarrow$
$X' = X[\langle Name, pid, t, locals, lChansOwned\rangle /$
$\langle Name, pid, s, locals, lChansOwned\rangle]$
$g' = g[Exclusive = \{\},$
$Handshake = \{\langle pid, ChanId(var, pid, \Sigma), [\![expr]\!]_{pid,\Sigma}\rangle\}]$
Note that the execution of a synchronous *send* statement in *atomic* mode causes the process to lose the right to execute the next statement (that right passes to the receiving process).
8. $(\alpha.label \equiv var?const \wedge capacity(var, pid, \Sigma) = 0$
$\wedge\ WillFail(\alpha.label, pid, \Sigma) = ff)$
$\wedge\ Handshake = \{\langle qid, [\![var_1]\!]_{pid,\Sigma}, const\rangle\}$
$\Longrightarrow$
$X' = X[\langle Name, pid, t, locals, lChansOwned\rangle /$
$\langle Name, pid, s, locals, lChansOwned\rangle]$
$g' = g[\mathbf{Exclusive} = \{pid\}, Handshake = \{\}]$

9. $(\alpha.label \equiv var_1?var_2 \wedge capacity(var_1, pid, \Sigma) = 0$
$\wedge\ WillFail(\alpha.label, pid, \Sigma) = ff)$
$\wedge\ Handshake = \{\langle qid, [\![var_1]\!]_{pid,\Sigma}, const\rangle\}$
$\Longrightarrow$
$X' = X[\langle Name, pid, t, locals, lChansOwned\rangle /$
$\langle Name, pid, s, locals', lChansOwned\rangle]$
$g' = g[\textbf{Exclusive} = \{pid\}, globals'/globals, ChIdMap'/ChIdMap]$
where

- $islnumvar(var_2) = tt$
  $\Longrightarrow (locals' = locals[var_2 = const], globals' = globals,$
  $ChIdMap' = ChIdMap)$
- $ischan(var_2) = tt$
  $\Longrightarrow (locals' = locals, globals' = globals,$
  $ChIdMap' = ChIdMap[ChanId(var_2, pid, \Sigma) = const]$
- $isgnumvar(var_2) = tt$
  $\Longrightarrow (globals' = globals[var_2 = const], locals' = locals,$
  $ChIdMap' = ChIdMap)$

10. $(\alpha.label \equiv run\ Name_1\ arg) \wedge WillFail(\alpha.label, pid, \Sigma) = ff) \wedge$
$\langle Name_1, Structure_1, Start_1, locals_1, ChansOwned_1, Active_1, Param_1\rangle$
is a symbolic transition system $\Longrightarrow$
$X" = X[\langle Name, pid, t, locals, lChansOwned\rangle /$
$\langle Name, pid, s, locals, lChansOwned\rangle]$
$X' = X'' \cup \langle Name_1, NrProcs + 1, Start_1, locals_2, ChansOwned_1\rangle$
$locals_2 = locals_1[Param_1 = [\![arg]\!]_{pid,\Sigma}]$
$g' = g[\textbf{Exclusive} = \{pid\}, NrProcs = NrProcs + 1,$
$Incorp(ChIdMap, ChansOwned_1, NrProcs + 1)/ChIdMap,$
$IncorpCont(ChContMap, ChIdMap, lChansOwned_1, NrProcs + 1)/$
$ChContMap, \textbf{dstep}'/\textbf{dstep}]$
where $\textbf{dstep}' = 1$ *if* $\alpha.label = deterministic$ *and* 0 *otherwise.*
11. $(\alpha.label \equiv end) \wedge WillFail(\alpha.label, pid, \Sigma) = ff) \wedge$
$\Longrightarrow$
$X' = X \setminus \{\langle Name, pid, t, locals, lChansOwned\rangle\}$
$g' = g[RetractId(ChIdMap, lChansOwned, pid)/ChIdMap,$
$RetractCont(ChContMap, ChIdMap, lChansOwned, pid)/ChContMap,$
$NrProcs = NrProcs - 1]$
12. $((\alpha.label \equiv var!expr) \vee (\alpha.label \equiv var?convar) \wedge$
$capacity(var, pid, \Sigma) = 0 \wedge \alpha.mode = deterministic$
$\Longrightarrow$
$\Sigma' = \Omega$
Any attempt to perform synchronous communication (rendezvous handshaking) in deterministic mode fails.

5. Executions

In this section, we define the executions of a PROMELA program. So far we have not taken into account the priorities of the transitions into considerations. We define an executable transition to be *selectable* if there exists no other executable statement with strictly higher priority. The following definition formalizes this statement.

DEFINITION 5.1. *A* selectable *transition of a state* $\Sigma = \langle X, g\rangle$ *is an executable transition of* $\langle pid, Name, s, \alpha, t\rangle$ *such that if* $\langle pid, s, \beta, t'\rangle$ *is any executable transition of* Σ *then* $\alpha.priority > \beta.priority$.

DEFINITION 5.2. *If a transition t is selectable in state Σ then* $\Sigma \rightarrow_1 Effect(\Sigma, t)$.

Selectable transitions almost correspond to actual execution steps of a program, but there are some steps that have not yet been modeled.

If, for instance, a process becomes blocked while it is executing an atomic sequence, then the process loses its exclusive control (Page 98 of [**Hol95**]). However, if a process becomes blocked when it is executing a *dstep* sequence (i.e., when it executes in *deterministic* mode), then it fails. We define $\rightarrow_2$-transitions to augment $\rightarrow_1$-transitions with these additional steps.

DEFINITION 5.3. $\rightarrow_2$ *is the smallest binary relation over Gstates satisfying the following conditions.*

1. $\Sigma \rightarrow_1 \Sigma' \implies \Sigma \rightarrow_2 \Sigma'$.
2. $(\langle X, g\rangle \neq \Omega \wedge \neg(\exists \langle X', g'\rangle : \langle X, g\rangle \rightarrow_1 \langle X', g'\rangle) \wedge \langle X, g\rangle[g.Exclusive = \{\}] \rightarrow_1 \Sigma'' \wedge g.dstep = 0) \implies \langle X, g\rangle \rightarrow_2 \Sigma''$.
3. $(\langle X, g\rangle \neq \Omega \wedge \neg(\exists \langle X', g'\rangle : \langle X, g\rangle \rightarrow_1 \langle X', g'\rangle) \wedge g.dstep = 1) \implies \langle X, g\rangle \rightarrow_2 \Omega$.

We have still not captured all the possible execution steps of a PROMELA program. The predefined variable *timeout* is set to be 1 when there are no executable statements in the system (Page 125 of [**Hol91**]). The executions that result by the setting of *timeout* to 1, augment $\rightarrow_2$-transitions with $\rightarrow_3$-transitions, as defined below.

DEFINITION 5.4. $\rightarrow_3$ *is the smallest binary relation over Gstates satisfying the following conditions.*

1. $\Sigma \rightarrow_2 \Sigma' \implies \Sigma \rightarrow_3 \Sigma'$.
2. $(\langle X, g\rangle \neq \Omega \wedge \neg(\exists \langle X', g'\rangle : \langle X, g\rangle \rightarrow_1 \langle X', g'\rangle) \wedge \langle X, g\rangle[g.timeout = 1] \rightarrow_2 \langle X'', g''\rangle) \implies \langle X, g\rangle \rightarrow_3 \langle X'', g''\rangle[g''.timeout = 0]$.

In these semantics we have modeled some conceptual single-step executions as multiple-step executions. For example, synchronous communication is modeled in two steps (a *send* followed by a *receive*). This resulted in the introduction of an "invisible" system state (the one in which the value is already sent but has not yet been received). Similarly, the intermediate states in a *dstep* sequence are invisible. The following definition formalizes this.

DEFINITION 5.5. *A state* $\Sigma = \langle X, g\rangle$ *is* invisible *if* $g.Handshake \neq \{\}$ *or* $g.Exclusive \neq \{\}$. *Any state that is not invisible is visible. The failure state* Ω *is always* visible. *Further* $\Sigma = \langle X, g\rangle$ *is* dinvisible *if* $g.dstep = 1$.

It is easy to show that every *dinvisible* state is also *invisible*. According to Page 98 of [**Hol95**], a dstep sequence is completely deterministic and any nondeterminism encountered may be resolved in an predetermined, but undefined way. The following definition covers this.

DEFINITION 5.6. $\rightarrow_4$ *is a largest subset of* $\rightarrow_3$ *such that for every* Σ_d*, a dinvisible state the following property holds:*

$$(\Sigma_d \rightarrow_4 \Sigma'_d \wedge \Sigma_d \rightarrow_4 \Sigma''_d) \implies \Sigma'_d = \Sigma''_d$$

Clearly $\rightarrow_4$ is not unique; this corresponds to the possibility of resolving the nondeterminism in more than one way, The only requirement we have on this relation is that it is the largest subset of $\rightarrow_3$ satisfying the above mentioned property.

We are now finally in a position to define the relation $\rightarrow$ which exactly defines the execution steps of a PROMELA program.

DEFINITION 5.7. *$\rightarrow$ is is the smallest binary relation over visible states defined as follows. If Σ and Σ' are visible global states then*

1. $\Sigma \rightarrow_4 \Sigma' \implies \Sigma \rightarrow \Sigma'$.
2. *If $\Sigma_1, \cdots, \Sigma_k$ is a finite sequence of invisible states for some $k > 0$ such that $\Sigma \rightarrow \Sigma_1$ and $\Sigma_k \rightarrow \Sigma'$ and $\Sigma_i \rightarrow \Sigma_{i+1}$ for $1 \leq i \leq k$, then $\Sigma \rightarrow \Sigma'$.*
3. *$\Sigma \rightarrow \Sigma$ if there exists an infinite sequence of invisible states $\{\Sigma_i\}$ and $\Sigma \rightarrow \Sigma_0$ and $\Sigma_i \rightarrow \Sigma_{i+1}$ for all $i \in$ Naturals.*

6. Example

In this section, we illustrate by means of an example how a process declaration in textual form is represented as a symbolic labeled transition system.

Consider the following process type which is some arbitrary collection of PROMELA statements.

EXAMPLE 6.1. `active proctype foo(int x)`

```
{
chan c1,c2;
chan c3 = [10] of {chan};
int x1 = 1;
int x2;
x1 = 2;
atomic { (x1 == 2) -> x1 = x1 + 1 }
do
::  (x1 > 5) -> x1 = x1 - 5
::  else -> x1 = x1 + 5
od;
skip;
c1?x1 unless c2?x2
}
```

The symbolic labeled transition system corresponding to this program is

$$\langle foo, Structure, q_0, locals, ChansOwned, 1, x\rangle$$

where *Structure* is given in Figure 1. *locals* is the partial function which takes values $0, 0, 1$ at $x, x2, x1$ respectively and undefined at all other arguments. *ChansOwned* is the finite sequence $\langle c1, \Omega, \Omega\rangle, \langle c2, \Omega, \Omega\rangle, \langle c3, 10, chan\rangle$.

The priority of the transition from state q_8 to q_{11} is higher than the one from q_8 to q_9. The *else* statement is translated into the negation of the conjunction of the other guards of the same *if* or *do* statement. It is an error to have *send* or *receive* statements as alternatives to an *else* in a *if* or *do* statement. Statements that transfer control out of an *atomic* into a non-atomic region are executed in *normal mode.*

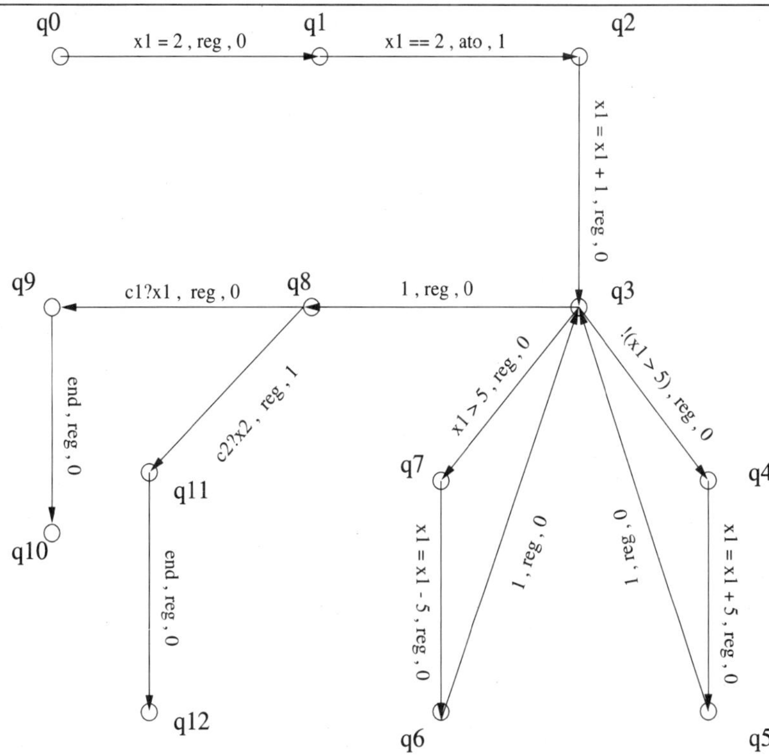

FIGURE 1. Sample Symbolic Labeled Transition System

7. Summary

This report gives an outline for an operational-semantics definition of the verification language PROMELA. We showed how a given PROMELA program, represented as sequence of symbolic labeled transition systems, can be translated into a Kripke structure. This was done by defining primitive statements as *conditional state transformers*. Not defined here is the conversion algorithm, part of SPIN, that translates PROMELA proctype definitions into symbolic labeled transition systems.

Features of the language that contribute to the complexity of the semantics include *mobility* (the ability to pass channel names over channels), and *atomic* and *deterministic* sequences. The first feature, for instance, necessitates the use of channel manipulation functions (Subsection 2.6.5) in the semantics of *run* and *end* statements.

For this outline we have made several simplifying assumptions about PROMELA syntax. For instance, we have not considered

- Correctness properties.
- Array variables.
- Data structures.
- Channels with more than one message field.
- Run statements with more than one argument.

The semantics of the *run* statement, furthermore, is somewhat different from that presented in Page 95 of [**Hol91**]. In the semantics given here, the *run* statement cannot be used in composite arithmetic expressions. The semantics definitions can be modified fairly straightforwardly to overcome these restrictions.

References

[Hol91] G.J. Holzmann, *Design and Validation of Computer Protocols*, Prentice Hall, 1991.
[Hol94] G.J. Holzmann, *What's New in* SPIN *Version 2*, Online document, SPIN distribution, 1994.
[Hol95] G.J. Holzmann, *The Verification of Concurrent Systems*, Course Notes, To appear.

NETWORKING HARDWARE DIVISION, IBM CORPORATION, RTP, NC, USA
E-mail address: `natarajraleigh.ibm.com`

BELL LABORATORIES, MURRAY HILL, NJ, 07974 USA
E-mail address: `gerardresearch.bell-labs.com`

DIMACS Series in Discrete Mathematics
and Theoretical Computer Science
Volume **32**, 1997

A Simulation and Validation Tool for Self-Stabilizing Protocols

Sandeep K. Shukla, Daniel J. Rosenkrantz, and S. S. Ravi

ABSTRACT. We discuss the development of a SPIN [**Hol90**] based tool that facilitates the simulation and validation *self-stabilizing* distributed protocols. In simple terms, our tool is a front-end that translates a language in which most self-stabilizing algorithms in the literature (e.g. [**Dij74, BP89, FD94, GK93, Her90, Hua93, SRR95**]) are described into PROMELA, the language handled by SPIN. Our input language is similar to Dijkstra's guarded command language [**Dij76**], with a slightly different semantics. One distinct feature of this input language is that it allows a user to specify the architecture (e.g. rings, complete graphs, linear arrays) of the system and other relevant information (uniform vs. nonuniform algorithms, oriented vs. unoriented links etc.). The main advantage provided by the tool is that it enables designers of self-stabilizing protocols to use the SPIN system to test and debug their protocols. Thus, early detection of bugs in an algorithm is facilitated. Our language can be extended to describe different architectures and to allow the specification of different types of faults and scheduling disciplines.

1. Self-Stabilizing Protocols

The concept of **self-stabilization**, introduced by Dijkstra [**Dij74**], has been of considerable interest to researchers in the area of fault-tolerant distributed systems. Self-stabilization provides a uniform approach to fault-tolerance [**Sch93**]. Due to transient faults or arbitrary initialization, a distributed system may enter an undesirable or **illegitimate** global state [**Dij74**]. In such situations, a self-stabilizing algorithm (protocol) enables the system to recover to a legitimate global state in a bounded amount

---

1991 *Mathematics Subject Classification.* Primary 68Q65, 68Q22, 68M15; Secondary 68N20.

Key words and phrases. Verification, Simulation, Distributed Systems, Self-Stabilization, Protocols, SPIN System.

This research was supported in part by NSF Grants CCR-94-06611 and CCR-90-06396.

of time. Self-stabilizing algorithms have been developed for a number of problems (see [**Sch93, EK89**] and the references cited therein).

A self-stabilizing protocol design problem is the problem of designing a distributed algorithm such that when the system stabilizes, the resulting global state is legitimate; that is, it satisfies a specified requirement. As an example, consider the problem of 2-coloring an even ring. Here, the processor configuration is an unoriented ring with an even number of processors. The requirement for a legitimate state is that all the nodes of the ring are colored either 1 or 0 and no two adjacent nodes have the same color. We are required to design a distributed algorithm that will restore the system state to a state where the colors assigned to the nodes satisfy the 2-coloring requirement whenever a transient fault disturbs the legitimate configuration into an illegitimate one. Moreover, the algorithm must enable the system to reach such a state from *any* initial state.

A distributed algorithm is **uniform** if each processor in the distributed system executes the same program. Uniform self-stabilizing algorithms (USSAs) are known for some problems including 6-coloring planar graphs [**GK93**], finding centers and medians for trees [**KPBG94**], 2-coloring certain rings and chains [**SRR94**], orienting odd-length rings [**Hoe94**], and leader election in rings of prime size [**Hua93**]. For some problems, it has been shown [**Ang80, BP89, IJ93**] that *deterministic* uniform self-stabilizing algorithms (DUSSAs) are impossible because of the difficulties encountered in deterministic **symmetry breaking**. For several such problems, researchers have presented randomized algorithms that self-stabilize with high probability; see for example, [**Her90, IJ93, SRR94**].

It is generally desirable to develop DUSSAs for problems on anonymous networks rather than non-uniform algorithms for ID-based networks. Unfortunately, for anonymous networks, DUSSAs are impossible even for very simple problems (e.g., 2-coloring an anonymous linear array of processors with an even number of nodes [**SRR94**]).

However, designing and proving the correctness of self-stabilizing protocols is a formidable task even for very simple problems. Usually, in the process of designing such protocols, hand simulation of a candidate solution on small problem instances helps in finding bugs in the design. However, it is a tedious task to try all possible arbitrary initializations in such simulations because the number of possibilities is often extensively large. This motivated us to develop a tool for early detection of bugs in a design via automated simulation and validation of instances.

Although simulation and validation tools for distributed protocols are currently available (e.g., SPIN [**Hol90**], Concurrency Factory [**CGL$^+$94**]), none of them supports direct simulation and validation of self-stabilizing protocols. Two essential features that a simulator for self-stabilizing protocols should have are arbitrary initialization of the processor states during simulation and the ability to describe the protocols at an abstract level without getting into implementation details specific to an architecture.

We have designed and implemented a simulation and validation tool for self-stabilizing protocols. This tool has been implemented on top of SPIN [**Hol90**], a widely available tool for simulating and validating distributed protocols. An important component of our work is a description language for self-stabilizing protocols which facilitates concise descriptions of protocols with architectural information. We define the semantics for such a language and also provide a translator from this language to the specification language PROMELA which is used for modeling protocols in SPIN. This front-end to SPIN achieves many of our objectives in creating a tool meant for simulation and validation of self-stabilizing protocols. Note that an expert user of SPIN may achieve this task by manually translating the protocol description into PROMELA. In order to do so, however, a user must possess a detailed knowledge of SPIN and PROMELA. Most of the self-stabilizing protocols presented in the literature have been based on a very simple model of a distributed system, namely, the shared memory model used in Dijkstra's original work [**Dij74**]. Further, the syntax used in describing self-stabilizing algorithms is very similar to the guarded command notation [**Dij76**]. Simple models and language constructs enable us to understand the central issues concerning the design of distributed algorithms for a problem. Tools that allow such models and descriptions would be valuable to researchers since algorithms can be presented to such tools in the same form as they appear in the literature. Therefore, allowing a syntax similar to guarded commands and allowing architectural specifications were assumed to be important requirements for our tool.

As mentioned earlier, one can find many examples in the literature on self-stabilizing protocols, where the algorithms are expressed in a language similar to the guarded command language [**Dij76, DF88**]; see for example, [**Dij74, GK93, Hua93, Her90, BP89, AG92, SRR94, SRR95, KPBG94, Sch93, AB93, BGM93, EK89, FD94, Hoe94**]. Further, many of the self-stabilizing distributed algorithms are designed for specific architectures; for example, self-stabilizing token ring [**Dij74**], self-stabilizing leader election in odd size rings [**Hua93**], self-stabilizing median finding in trees [**KPBG94**], self-stabilizing coloring and other problems for rings, chains and bipartite graphs [**SRR94, SRR95**]. We note that the description language used by the research community in presenting these protocols is at a level of abstraction where there is no need to include details concerning data structures and architectural parameters. In general, the architecture for which the protocol is designed is mentioned separately and the protocol is presented at an abstract level. Thus, a simulator for such protocols should be able to accept a description or model of the protocol which is at a similar level of abstraction.

Moreover, if the syntax of the input language to the simulator is similar to the syntax of the language used in describing these algorithms and there is a facility for describing the architecture for which the algorithm is designed, then a direct simulation of the protocols described in an abstract level can be supported. This is useful in testing and debugging via simulation and making validation models for these distributed algorithms which are otherwise prone to subtle errors.

2. A Description of the Tool

Here, we describe the input language, called **gcap** ("**g**uarded **c**ommand with **a**rchitecture **p**rimitives") for our tool and a prototype implementation of a translator for **gcap** into PROMELA. Since PROMELA is the modeling language for SPIN, once translation into PROMELA is accomplished, one can directly use SPIN for simulating and validating. To the best of our knowledge, none of the widely available tools for simulation of distributed protocols has the facility required for simulating and validating self-stabilizing protocols directly.

In achieving our objective, we have defined the semantics of **gcap** in terms of PROMELA. A detailed description of **Promela** (PROcess MEta LAnguage) can be found in [**Hol90, Hol93**]. Our translator serves as a front-end to this simulator so that a user can specify architecture dependent protocols using a simple input language. Further, our translator creates a random initial state for the system for each separate compilation.

Consider, for example, Dijkstra's [**Dij74**] self-stabilizing token ring protocol. A PROMELA specification of that protocol (for a ring of size 10) is shown in Appendix A. For arbitrary initialization, one has to manually change the initialization part of the PROMELA code every time. Moreover, to make sure that the appropriate architecture is assumed by the simulator, a detailed design of the global data structure has been made. One can easily see that the local state of each process is simulated as a position in a global array indexed by the process numbers. The variable *state* in each process is now called *state[index],* and when a process refers to a state variable of its left neighbor, it actually refers to *state[(index-1)mod size]* where *size* is the size of the ring. Thus, a description of this protocol in PROMELA requires a user to pay attention to low level architectural details. Also, PROMELA requires the specification of an *init* process that starts the execution of all the other processes. In other words, the simplicity provided by an abstract specification of the protocol is lost and a user is forced to consider issues (such as the maintenance of global data structures) that are specific to modeling in PROMELA rather than the issues central to the correctness of the protocol. Hence, a language that allows a user to specify protocols at a higher level of abstraction is easier to use. For example, the protocol in Appendix A can be specified in **gcap** as follows.

```
UNRING(10,P,Q);
PROCESS P{
  byte state;
  $< state = state(left) >$ ->
        $<state = state +1 mod 10 >$;
}
```

```
PROCESS Q{
  $< state != state(left) >$ ->
        $<state = state(left) >$;
}
```

We now explain the constructs used in the above specification. The keyword `UNRING` specifies that the architecture on which the protocol is supposed to run is a U*nidirectional* `RING` and the protocol is N*on-uniform*. A non-uniform algorithm is described by specifying a process to be executed by a distinguished *leader* processor and another process to be run on the rest of the processors. The keyword `UNRING` requires requires three parameters. The first parameter (`10` in the above example) gives the number of processors in the ring. The second and third parameters (`P` and `Q`) specify the names of the processes to be run on the leader and the other processors respectively. The description for a process is a collection of guarded commands. Each guarded command contains a guard (a Boolean condition on the states of the executing processor and other processors that can directly communicate with the executing processor as determined by the processor topology) and an action (a sequence of statements). Syntactically, each guard and action are bracketed by "`$<`" and "`>$`". The semantics of the sequence of guarded command is that a nondeterministic choice of a command is made between from the set of guarded commands which have their guards true when the process is scheduled; and the action corresponding to the guarded command is executed atomically.

Our translator takes this specification and generates low level detailed Promela code specific to the Spin simulator. Moreover, our translator uses a random number generator to achieve arbitrary initialization in the *init* process in each separate compilation. Hence, to simulate distinct arbitrary initializations, one can compile the code several times, and the initialization will be different with high probability. We have developed an LL(1) grammar for **gcap**. The translator uses resursive descent parsing and standard syntax-dircted translation methods to produce the Promela specification.

The current version of the language supports simple architectures such as oriented/unoriented rings, chains etc. However, it is relatively straightforward to add syntax that will enable us to specify other architectures and modify the translator accordingly.

3. Extensions for Validating Self-Stabilizing Protocols

Most of the research papers on self-stabilizing protocols describe protocols that are parameterized by the number of processors. Hence a correctness proof needs to be carried out without relying on a specific number of processors. To automate such correctness proofs, one might use theorem provers but that would require a lot of user interaction. However, from an implementation point of view, it might be useful to prove the correctness for a specific number of processors. This specific number may correspond to number of processors on which the system is being implemented.

Suppose that legitimate global states are specified by a formula f. In order to validate that an instance of an algorithm self-stabilizes of an instance of the algorithm, we want to prove that "always eventually f" holds and once f holds, none of the possible transitions can lead to a state where f does not hold. In the language of linear time temporal logic (LTL) this can be expressed as $GFf \wedge G(f \rightarrow Gf)$. Our front-end can be easily extended to accept the predicate that characterizes legitimacy and then to generate the LTL formula as above. lso The front-end compiler will insert a **never** claim in the PROMELA file that will be submitted to SPIN.

We are also considering some inductive proof methods similar to [**KM89, Kur94**] which will help us to prove the correctness for any number of processors.

Another extension that we have planned is to allow the specification of a scheduling adversary in **gcap**. The user can specify what kind of adversary is assumed (central scheduler, distributed scheduler or other stronger adversaries). The front-end compiler will generate different PROMELA code for each case.

Another important advantage of using SPIN as the core is that the underlying process description language PROMELA allows a user to model several forms of faults within the language in a manner similar to [**Aro93**]. Thus, a useful extension of of the input language is to allow the specification of a transient fault model along with the algorithm. This will provide a facility for experimenting with different types of transient faults.

4. Concluding Remarks

The detailed syntax and semantics of our description language will appear in a future version of this paper. In general, the syntax is close to the informal algorithmic description languages used for presenting self-stabilizing protocols in the literature. Details regarding the translation procedure, architectural specifications supported currently and future enhancements will be included in a future version of the paper. In Appendix B we have included a number of examples of self-stabilizing protocols from [**Dij74, SRR94, SRR95**] described in **gcap** and the corresponding PROMELA specifications generated by our translator. It should be noted that the current version of our translator does not attempt to optimize the PROMELA code. For example, instead of using `init` on 10 different processes we could have used a PROMELA feature `active [10] proctype A()` which also creates 10 instances. In the next version of the tool we plan to incorporate such optimizations.

Acknowledgement: We wish to thank Professor Mohamed Gouda (University of Texas at Austin) for interesting discussions and Dr. Gerard Holzmann (Bell Labs.) for valuable comments on a previous version of this paper.

References

[AB93] Y. Afek and G. M. Brown, *Self-stabilization over unreliable communication media*, Distributed Computing **7** (1993), 27–34.

[AG92] M.S. Abadir and M.G. Gouda, *The stabilizing computer*, Proceedings of the 1992 international conference on Parallel and Distributed systems., 1992.

[Ang80] D. Angluin, *Local and Global properties in networks of processes*, Proceedings of the 12th Annual ACM symposium on theory of computing (University of California at Los Angeles), ACM, 1980.

[Aro93] Anish Arora, *A foundation of fault tolerant computing*, PhD Thesis, University of Texas at Austin, 1993.

[BGM93] J. E. Burns, M. G. Gouda, and R. E. Miller, *Stabilization and pseudo-stabilization*, Distributed Computing **7** (1993), 35–42.

[BP89] J. E. Burns and J. Pachl, *Uniform self-stabilizing rings*, ACM Transactions on Programming Languages and Systems **11** (1989), no. 2, 330–344.

[CGL+94] R. Cleaveland, J. N. Gada, P. M. Lewis, S. A. Smolka, O. Sokolsky, and S. Zhang, *The Concurrency Factory– practical tools for specification, simulation, verification and implementation of concurrent systems*, DIMACS workshop on Specification of Parallel Algorithms, 1994.

[DF88] E. W. Dijkstra and W. H. J. Feijen, *A method of programming*, Addison-Wesley Publishing Co., 1988.

[Dij74] E. W. Dijkstra, *Self-stabilizing systems inspite of distributed control*, Communications of ACM **17** (1974), no. 11, 643–644.

[Dij76] E. W. Dijkstra, , Prentice Hall Series in Automatic Computation, Prentice Hall, Inc., 1976.

[EK89] M. Evangelist and S. Katz (eds.), *Proceedings of the mcc workshop on self-stabilizing systems*, MCC, Austin, TX, 1989.

[FD94] M. Flatebo and A. K. Datta, *Two-state self-stabilizing algorithms for token rings*, IEEE Transactions on Software Engineering (1994).

[GK93] S. Ghosh and M. H. Karaata, *A self-stabilizing algorithm for coloring planar graphs*, Distributed Computing **7** (1993), no. 1, 55–59.

[Her90] T. Herman, *Probabilistic self-stabilization*, Information Processing Letters **35** (1990), no. 2, 63–67.

[Hoe94] J.-H. Hoepman, *Uniform deterministic self-stabilizing ring-orientation on odd-length rings*, Tech. Report CS-R9423, Computer Science Department, CWI, Amsterdam, April 1994.

[Hol90] G. J. Holzmann, *Design and validation of computer protocols*, Prentice Hall Software series, Engelwood Cliffs, NJ, 1990.

[Hol93] G. J. Holzmann, *Design and validation of protocols: A tutorial*, Computer Networks and ISDN Systems **25** (1993), no. 9, 981–1017.

[Hua93] S. T. Huang, *Leader election in uniform rings*, ACM Transactions on Programming Languages and Systems **15** (1993), no. 3, 563–573.

[IJ93] J. Israeli and M. Jalfon, *Self-stabilizing ring orientation*, Information and Computation **104** (1993), no. 2, 175–196.

[KM89] R. P. Kurshan and K. P. McMillan, *A structural induction theorem for processes*, Proceedings of Conference on Principles of Distributed Computing (PODC), ACM, 1989, pp. 239–247.

[KPBG94] M.H. Kaarata, S.V. Pemmaraju, S.C. Bruell, and S Ghosh, *Self-stabilizing Algorithms for Finding Centers and Medians of Trees*, TR- 94-03, University of Iowa, 1994.

[Kur94] R. Kurshan, Automata Theoretic Verification of Coordinating Processes, Princeton Series in Computer Science, Princeton University Press, 1994.

[Sch93] M. Schneider, *Self-stabilization*, Computing Surveys **25** (1993), no. 1, 45–67.

[SRR94] S. K. Shukla, D.J. Rosenkrantz, and S.S. Ravi, *Developing Self-Stabilizing Coloring Algorithms via Systematic Randomization*, Proceedings of the 1st International Workshop on Parallel Processing (New Delhi), Tata McGraw-Hill, 1994, pp. 668–673.

[SRR95] S. K. Shukla, D.J. Rosenkrantz, and S.S. Ravi, *Observations on self-stabilizing graph algorithms*, 2nd Workshop on Self-stabilizing Systems (Las Vegas), University of Nevada at Las Vegas Tech. Report, 1995.

Appendix A. Dijkstra's Token Ring Protocol in Promela

```
#define N  10
#define size 10

byte state[size];

proctype A(short index)
{   do
    :: atomic{ state[index] == state[N-1] ->
           if
           :: (state[index] + 1 == 10) -> state[index] = 0
           :: (state[index] + 1 < 10) ->
                      state[index] = state[index] +1
           fi
           }
    od
}
proctype B(short index)
{  do
   :: atomic{ (state[index] != state[index -1]) ->
                      state[index] = state[index -1]
           }
   od
}
init
{  atomic{ state[0] = 1; state[1] = 2; state[2] = 3;
          state[4] = 4; state[5] = 5; state[6] = 3;
          state[7] = 4; state[8] = 6; state[9] = 9};
  atomic{ run A(0); run B(1); run B(2); run B(3); run B(4);
        run B(5); run B(6); run B(7); run B(8); run B(9)}
}
```

Appendix B. Examples of Translations

B.1. Dijkstra's Token Ring on a 10 processor unidirectional Ring. The **gcap** specification of Dijkstra's Token Ring algorithm to be run on a 10 processor unidirectional nonuniform ring is given below.

```
RING(10,P,Q);
int state;
PROCESS P{
  $<state = state(left) >$ -> $<state := state + 1 >$
}
PROCESS Q{
  $<state != state(left) >$ -> $<state := state(left) >$
}
```

Here is the PROMELA code generated by our translator:

```
#define size 10
int state[10] ;

proctype P(int index)
{ do
  :: atomic{
     state[index] == state[(size + index - 1) % size] ->
            state[index] = state[index] + 1 }
  od
}
proctype Q(int index)
{ do
  :: atomic{
     state[index] != state[(size + index - 1) % size] ->
            state[index] = state[(size + index - 1) % size] }
  od
}
init
{ atomic{
  state[0] = 2; state[1] = 3; state[2] = 0; state[3] = 1;
  state[4] = 2; state[5] = 3; state[6] = 0; state[7] = 1;
  state[8] = 2; state[9] = 3;
  };
  atomic{
  run P(0); run Q(1); run Q(2); run Q(3); run Q(4);
  run Q(5); run Q(6); run Q(7); run Q(8); run Q(9);
  }
}
```

B.2. Maximal Independent Set Algorithm on a Bidirectional Uniform Ring of size 10. [**SRR95**]. Here is the **gcap** specification of thealgorithm to be simulated.

```
RING(10,P,Q);
bit Ind;
```

```
PROCESS P{
$< (Ind(left) = 0) && (Ind(right) = 0) && (Ind = 0) >$ ->
            $< Ind := 1 >$
|
$< ((Ind(left) = 1) || (Ind(right) = 1)) && (Ind = 1) >$ ->
            $<Ind := 0 >$
}

PROCESS Q{
$< (Ind(left) = 0) && (Ind(right) = 0) && (Ind = 0) >$ ->
            $< Ind := 1 >$
|
$< ((Ind(left) = 1) || (Ind(right) = 1)) && (Ind = 1) >$ ->
            $<Ind := 0 >$
}
```

Here is the PROMELA specification generated by our system:

```
#define size 10
bit Ind[10] ;

proctype P(int index)
{ do
  :: atomic{
     ( Ind[(size + index - 1) % size] == 0 ) &&
        ( Ind[(index + 1) % size] == 0 )
     && ( Ind[index] == 0 ) -> Ind[index] = 1 }
  :: atomic{
     ( ( Ind[(size + index - 1) % size] == 1 ) ||
        ( Ind[(index + 1) % size] == 1 ) )
     && ( Ind[index] == 1 ) -> Ind[index] = 0 }
  od
}
proctype Q(int index)
{ do
  :: atomic{
     ( Ind[(size + index - 1) % size] == 0 ) &&
        ( Ind[(index + 1) % size] == 0 )
     && ( Ind[index] == 0 ) -> Ind[index] = 1 }
  :: atomic{
     ( ( Ind[(size + index - 1) % size] == 1 ) ||
       ( Ind[(index + 1) % size] == 1 ) )
    && ( Ind[index] == 1 ) -> Ind[index] = 0 }
  od
}

init
```

```
{ atomic{
    Ind[0] = 1; Ind[1] = 1; Ind[2] = 0; Ind[3] = 0; Ind[4] = 1;
    Ind[5] = 1; Ind[6] = 0; Ind[7] = 0; Ind[8] = 1; Ind[9] = 1;
    };
    atomic{
    run P(0); run Q(1); run Q(2); run Q(3); run Q(4);
    run Q(5); run Q(6); run Q(7); run Q(8); run Q(9);
    }
}
```

B.3. Coloring of Odd chains on a chain of Size 11 [**SRR95**]. Here is the **gcap** specification.

```
CHAIN(11, P, Q);
byte X, color;

PROCESS P{
$< X != 0 || color(index) != 0  >$ -> $<X := 0; color := 0 >$
}

PROCESS Q{
$< ((X(left) >= X(right)) && (X != X(right) + 1)) ||
                    ( color != X %2)  >$ ->
                     $<X := X(right) + 1; color := X %2 >$
|
$< ((X(left) < X(right)) && (X != X(left) + 1)) ||
                    ( color := X %2) >$ ->
                     $<X := X(left) + 1; color := X %2 >$
}
```

Here is the PROMELA Specification generated by our system:

```
#define size 11
byte X[11] , color[11] ;

proctype P(int index)
{ do
  :: atomic{  X[index] != 0 || color[index] != 0 ->
                      X[index] = 0; color[index] = 0 }
  od
}
proctype Q(int index)
{ do
  :: atomic{
      ( ( X[(index -1)] >= X[(index + 1)] ) &&
        ( X[index] != X[(index + 1)] + 1 ) ) ||
        ( color[index] != X[index] % 2 ) ->
    X[index] = X[(index + 1)] + 1;
```

```
    color[index] = X[index] % 2 }
  :: atomic{
     ( ( X[(index -1)] < X[(index + 1)] ) &&
       ( X[index] != X[(index -1)] + 1 ) ) ||
       ( color[index] = X[index] % 2 ) ->
     X[index] = X[(index -1)] + 1;
     color[index] = X[index] % 2 }
  od
}
init
{ atomic{
  X[0] = 0; X[1] = 0; X[2] = 4; X[3] = 1; X[4] = 4;
  X[5] = 4;  X[6] = 2; X[7] = 0; X[8] = 1; X[9] = 2;
  X[10] = 3;
  color[0] = 1; color[1] = 1; color[2] = 1; color[3] = 3;
  color[4] = 1; color[5] = 1; color[6] = 4; color[7] = 2;
  color[8] = 4; color[9] = 2; color[10] = 0;
  };
  atomic{
  run P(0); run P(10);run Q(1); run Q(2); run Q(3);
  run Q(4); run Q(5); run Q(6); run Q(7); run Q(8);
  run Q(9);
  }
}
```

SANDEEP K. SHUKLA, DEPARTMENT OF COMPUTER SCIENCE, UNIVERSITY AT ALBANY – STATE UNIVERSITY OF NEW YORK, ALBANY, NY 12222, USA.
E-mail address: sandeepcs.albany.edu

DANIEL J. ROSENKRANTZ, DEPARTMENT OF COMPUTER SCIENCE, UNIVERSITY AT ALBANY – STATE UNIVERSITY OF NEW YORK, ALBANY, NY 12222, USA.
E-mail address: djrcs.albany.edu

S. S. RAVI, DEPARTMENT OF COMPUTER SCIENCE, UNIVERSITY AT ALBANY – STATE UNIVERSITY OF NEW YORK, ALBANY, NY 12222, USA.
E-mail address: ravics.albany.edu

DIMACS Series in Discrete Mathematics
and Theoretical Computer Science
Volume **32**, 1997

Dynamic Analysis of SA/RT Models Using SPIN and Modular Verification

Javier Tuya, José R. de Diego, Claudio de la Riva, and José A. Corrales

ABSTRACT. We present a practical approach to use the SPIN model checker in conjunction with Structured Analysis methods for Real Time systems (SA/RT). The graphical model is translated into a PROMELA program in which we prove assertions about the desired behavior of the system. We also provide support for modular verification, by separately verifying different components of the model and deducing the desired global properties from the previous verifications. This approach is then illustrated using a steam boiler system as a case study.

1. Introduction

The development of reactive and real time systems is a critical task that requires our best efforts and techniques in order to attain more robust and reliable systems. The use of formal methods and verification techniques can help us in this task. We can distinguish between two distinct approaches for verification: theorem proving and model checking, but, from a practical point of view (and with the medium software engineer in mind), the use of model checking using the appropriate tools such as SPIN [**Ho91**] is simpler than that of theorem proving. Despite the state-explosion problems, the ability of model checkers to find bugs is impressive, making it easy to see the payoff from formal verification [**DR96**].

One of the more important aspects to consider in the development of real time systems is the use of adequate notations and tools [**CGR93**], which must be easy to use and must to generate documentation easy to review. Thus, our effort is aimed at the use of a dual language approach [**FMM94**], using graphical tools (to support the development of an operational specification) and model checkers (in order to prove properties for the operational model) as shown in Figure 1. Felder [**FMM94**] cites a large number of references in which the dual language approach is used. Others, in which the operational specification is developed using well known graphical methods, can be found in [**D93**], [**D94**], where Statecharts

1991 *Mathematics Subject Classification.* Primary 68Q60, 68Q45; Secondary 68U07, 93C83.
Key words and phrases. SPIN, PROMELA, Structured Analysis, SA/RT, Model Checking, Modular Verification.
This work has been funded by the "II Plan de Investigación del Principado de Asturias" (Spain) under project AUTOVER (PB-TIC95-01).

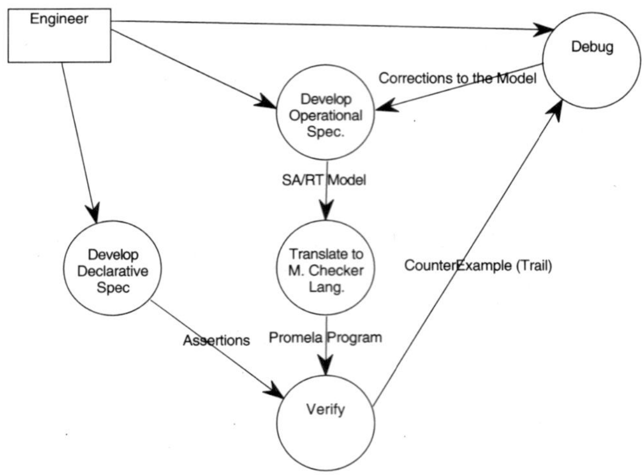

FIGURE 1. Dual language approach

are used in combination with a BDD based model checker; also in [**T94**], [**T95**] with Structured Analysis methods and the SMV model checker; and in [**L93**], [**A95**] where structured methods are used on top of a kernel of Petri nets that can be further used for verification purposes; or in [**LL95**], [**LL96**] where Message Sequence Charts (MSCs) are used in combination with SPIN.

In this paper, we will use Structured Analysis methods (SA/RT) to model the system. The remainder of this section gives an overview of the RSML semantics that we will use. In Section 2, we will show how we can translate the model into PROMELA source code and check it using the verifier generated by SPIN. Our approach is based on a two step execution semantics (micro-step and macro-step) and allows us to perform modular verification. In the second part of our paper we present a case study (Section 3) and show the use of SPIN for verification (Section 4).

1.1. SA/RT methods. Among the graphical methods most commonly used in industry, two of the leading methods are SA/RT and Statecharts. SA/RT is a short name for Structured Analysis methods with extensions for Real Time [**WM85**] [**HP87**]. The model is represented as a hierarchical set of diagrams that includes data and control transformations (processes). Control transformations are specified using State Transition Diagrams (STDs), and events are represented using control flows. The other graphical and state based formalism for specification of real time systems is Statecharts [**Ha87a**], [**Ha90**]. The system is represented as a set of hierarchical states instead of processes. Each state can be decomposed into substates and so on. The statecharts notation is more compact than the SA/RT notation and has been formally defined. If this is so, why have we not selected Statecharts? The answer is simple:

- There are a lot of tools that support SA (both for Information Systems and Real Time), ranging from low to high cost, and these are well known by engineers. In contrast, there exist few (and expensive) tools for Statecharts.
- There is a lack of verification tools for SA methods, reduced in practice to syntax checking and completeness checking (e.g. balancing). The possibility to perform semantic checks will be a great added value to these tools.

1.2. Step semantics. Before carrying out any verification effort, a clear execution semantics must be defined for the model. The original (informal) definition provided by Ward is inspired by the execution rules of Petri nets. The execution of the model consists of a series of execution steps (external interactions). At each one, the environment can evolve, by placing some token(s) into flows. The system responds by executing some transformation(s) that place tokens into output flows and removing the input tokens. If some tokens are placed in control flows that can activate other transitions, the interaction continues until no transition can be executed. Then, other execution steps can start. This execution semantics has a high degree of non-determinism, because when two (or more) tokens are simultaneously placed, two (or more) branches of internal interactions must be executed sequentially, but in an arbitrary order. Each one must be completed before returning to the next [**W86**].

Another approach is based on the concept of macro-step and micro-step in Statecharts (their equivalent in Ward semantics as described above are external and internal interactions) [**Ha87b**]. At each macro-step, the environment changes and the system responds to it in a synchronous way, i.e. all inputs are taken simultaneously, the system responds in zero time and sets all outputs simultaneously. Each macro-step is composed of a chain of internal steps (micro-steps), in which a set of transitions are selected and executed in parallel. The new configuration can cause the execution of other transitions until no transition can be made. The essential difference with regard to Ward's approach is that more than one transition can be executed in parallel.

Although we have selected SA/RT syntax, mainly due to the availability of CASE tools, we use the Statecharts semantics because it reduces the non-determinism of the specifications and enforces a more rigorous causal ordering of the transitions taken within a step. Several different semantics have been defined for Statecharts [**Hu91**], always based on the concept of micro and macro-steps. We have adopted the more recent Statecharts semantics, used in the RSML language that can be found in [**L94**]. This is outlined below with a somewhat different notation.

Let EI be the set of External Input flows, EO the set of External Output flows, F the set of internal flows, and S the set of variables that defines the local state for each process. We denote by EI_i, EO_i, F_i, S_i the values of these variables at each step. The configuration C_i at step i is the set $EI_i \cup EO_i \cup F_i \cup S_i$. A macro-step starts due to a change in the environment variables that will have the set of values EI_0 and will give the new configuration C_0. In response, the system selects a set of transitions T_i to be triggered and executes them. It changes the current configuration from C_0 to C_1. The process repeats by triggering other transitions based on the previous configuration until no more transitions can be carried out. This execution is outlined below:

```
do (each macro-step)
  i=0
  Changes to the environment (new configuration is C0)
  do (each micro-step)
    Select Ti for current configuration Ci
    Execute transitions in Ti (new configuration is Ci+1)
    i=i+1
  until T=empty
end do
```

2. Using SPIN and PROMELA for verification

PROMELA provides a powerful mechanism for representing concurrent processes and the communication between them. But when using the RSML semantics, we must implement the simultaneous execution of several transitions and differentiate between micro and macro-steps. A direct translation from processes in the graphical model into processes in PROMELA will need a tight coordination between them. So, it is easier to implement the whole model as an unique process and the sequences of steps as loops.

2.1. Step semantics implementation with PROMELA. The preliminary layout of the PROMELA program that implements this semantics will be:

```
do
  Changes to the environment                    (1)
  do
    Select transitions T to be executed         (2)
    Reset input flows                           (3)
    if :: T<>empty -> Execute transitions       (4)
       :: else -> goto end_macro_step
    fi
  od
end_macro_step:
  Assert desired properties                     (5)
  Reset output flows
od
```

1. External input flows (EI) can change their values. Since our desire is to prove properties that must be true for any execution of the system, these values are non-deterministically selected. This is a more realistic (and difficult to verify) assumption than the ones that suppose some kind of model for the environment behavior.
2. Because the SPIN structure is sequential, we cannot execute a transition that modifies the output flows because its result could influence another transition at the same micro-step. So, our first step is to select the set of transitions to be executed. A hidden variable is used to determine if a transition will be executed.
3. Before executing selected transitions T, we set all input control flows in $F \cup EI$ to zero. This situation will be needed as a prerequisite of the next micro-step.

4. When a transition can be executed, we continue by executing it (performing the actions that modify EO, F and S).
5. When no transition can be executed, we have finished the macro-step, The resulting configuration C_i represents the action of the whole macro-step, and we can check the assertions about the desired properties.

2.2. A more refined layout. The above layout is intuitive, but not practical when implemented in PROMELA. It must be modified by considering the following observations:

- It generates a high number of states that are not relevant for our verification purposes. We wish to enclose all sentences that are executed in a macro-step in an `atomic` sentence. So, the number of stored states will be the same of the observable states at the end of a step.
- If we put a loop inside the `atomic` sentence, there will be a potential infinite loop that can cause the verifier to hang. This problem will be even greater when we further include the concept of ghost variables. So, the specification must be restructured in such a way that all sentences in a micro-step will be executed in an atomic sentence, and so, a macro-step will be composed of one or more atomic steps. We differentiate between micro and macro-steps using a variable named `MicroStep`.
- The properties we wish to prove will be stated using assertions about current values of variables at the end of each macro-step. Typical assertions must ensure response or safety properties that relate external input flows to external output flows and to states. Sometimes we need to make assertions about previous values of variables and about internal flows. So as, to be able to do this, we define the following simple predicates:
  1. $WAS0(f)$, where $f \in EI \cup F$: true if f was true at some instant in some preceding micro-ste p of the current macro-step.
  2. $WAS1(f)$, where $f \in EI \cup EO \cup F$: true if f was true at some instant of the preceeding macro-step.
  3. $INTERVAL(f_1, f_2)$, where f_1 , $f_2 \in EI \cup EO \cup F$: set to true when f_1 becomes true, and changed to false when f_2 becomes false.
- Data transformations must be treated differently to control transformations, because they can execute some user supplied code. Data flows are discretized to relevant values, having a special value *NotPresent* that determines its presence or absence. The code for a discrete transformation will be executed when data becomes available. A continuous data transformation has an associated variable that can have one of the following values: *Disabled*, *Enabled*, *Scheduled*. At the start of each macro-step all enabled continuous transformations are set to *Scheduled*, and so will be executed at the first micro-step. Execution leads the value of this variable to *Enabled*. If as a consequence of the execution of a transition a data transformation is enabled, its value is set to *Scheduled* and hence, executed at the next micro-step.

Taking into account the above considerations, the new schema for our PROMELA program will be:

```
do
  atomic {
    if :: MicroStep==0 ->
            Changes to the environment
            Calculate predicates on EI
       :: else -> skip
    fi;
    Select transitions to be executed
    Execute data transformations (if any)
    Reset input flows
    if :: Some transition will be executed ->
            Execute transitions
            Calculate predicates on F
            MicroStep=1
       :: else ->
            Assert desired properties
            Calculate oredicates on EO
            Reset output flows
            MicroStep=0
    fi;
  }
od
```

2.3. Support for modular verification. Verification of realistically sized systems is still an open problem due to the state-explosion problem. Although SPIN is very efficient, and enclosing each micro-step into an atomic sentence drastically reduces the set of reachable states, the problem still exists. Therefore, we will adopt a modular approach (also named rely/guarantee or assumption/commitment), in which properties will be stated for part of the model, and using the Abadi/Lamport composition theorem [**AL93**] we will deduce the truth of the property for the whole model. This theorem is based on the following induction rule:

$$\frac{\begin{array}{ccc} M_1 \models_{\phi_{E_1}} \phi_1 & , & M_2 \models_{\phi_{E_2}} \phi_2 \\ M_1 \models \phi_{E_2} & , & M_2 \models \phi_{E_1} \end{array}}{[M_1 || M_2] \models (\phi_1 \wedge \phi_2)}$$

Let M be the whole model, which can be decomposed into the parallel composition of $[M_1||M_2]$, and ϕ the desired safety property, which can be decomposed as $\phi = \phi_1 \wedge \phi_2$. In order to prove ϕ over M, it is sufficient to prove:

1. Each ϕ_i is true for the components M_i under the assumption ϕ_{E_i} for the environment of each M_i.
2. Each component M_i grants the assumption $\phi_{E_j}, i \neq j$ of each component over its environment (commitment).

Let us take a simple example represented in Figure 2. We wish to prove a property ϕ that relates EI_1 and EO_2. The model is divided into two components M_1 and M_2 communicated by internal flows fg_1, fg_2. The property ϕ is decomposed into two (both for M_1 and M_2 respectively). When we prove the property for M_2 we must assume some behavior for fg_2.

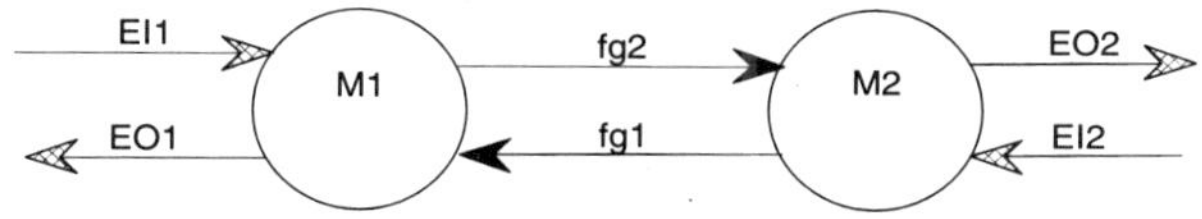

FIGURE 2. Modular verification

Internal variables (fg_1 and fg_2 in the example) that communicate components which will be separately verified will be named ghost variables. When the component M_2 is translated into PROMELA, we give its ghost variables (fg_2) random values. Since in the whole model they are internal variables, their values can change at any instant of any micro-step and the set of sentences that give values to ghost variables will be placed inside the corresponding portion of code.

Assumptions restrict the behavior of ghost variables, and so, we use a simple statement like:

$$ASSUME(cond, variable, value)$$

that must be read: If *cond* is true, then *variable* will be set to *value*. The following PROMELA code is placed immediately after the ghost variables have taken a value:

```
if :: cond -> variable=value
   :: else -> skip fi;
```

When we discharge the assumptions (i.e., prove the commitments for M_1), *ASSUME* sentences are coded as *COMMIT* sentences:

$$COMMIT(cond, variable, value)$$

which are then translated into a PROMELA assertion $cond \rightarrow variable = value$:

```
assert( ! cond || variable==value )
```

COMMIT sentences are translated and verified at the end of each micro-step, as opposed to model properties, which are checked only at the end of each macro-step.

By taking all the above considerations into account, the complete body of the PROMELA program will be:

```
          Declare variables and macros
          bit MicroStep==0
          active proctype SART()
          do
            atomic {
              if :: MicroStep==0 ->
progress:             Changes to the environment
                      Calculate predicates on EI
                 :: else -> skip
              fi;
              Select transitions to be executed
              Execute data transformations (if any)
              Reset input flows
              Set ghost variables
              Make assumptions
```

```
        if :: Some transition will be executed ->
                 Execute transitions
                 Assert commitments
                 Calculate predicates on F
                 MicroStep=1
           :: else ->
                 Assert desired properties
                 Calculate predicates on EO
                 Reset output flows
                 MicroStep=0
        fi;
      }          /* end atomic */
    od
```

2.4. Implementation issues. The automatic verification is currently being done as part of the AUTOVER project. The graphical specification (taken from a commercial CASE tool) is extracted to a textual file that describes the full set of transitions and variables using some C-like macros. For instance, `EXTINPUT(e)` represents an external input, `INTOUTPUT(f)` represents an internal flow, `GHOST(f)` establishes that f is a ghost variable, `TRANS(orig,dest,cond,acc)` represents a transition, `ASSUME(cond,f,v)` represents an assumption, etc.

The body of the PROMELA program is the same for all models and follows the structure outlined above. Each line of it is a section. At each section, and using standard preprocessor directives, the full model is processed, but using different definitions for the C macros. For instance, `TRANS` is defined as an assignment sentence that sets the value of one variable in T when included in the select transitions section, as a set of assignments for output flows when defined in the execute section, and as a declaration of a hidden variable (`T`) when included in the declarations section.

Assertions that are violated give us an execution trail that shows all the steps taken by the program until a specific assertion is violated. As we are only interested in the changes of the configuration and not in the internal steps taken by the SPIN program, we do not use the `-p` option, but timely print statements are placed in order to visualize the desired steps when following the trail (option `-t`). These print statements are coded as a macro that is defined as `skip` when we run the verifier and into `printf` when we wish to visualize the counter-example (a sample can be found in Section 4.2). On some occasions, counter-examples are extremely long. Due to this, we generate the verifier using the `-DREDUCE` option, and execute it with the `-i` or `-I` flag.

The experiments that will be presented in the following sections have been carried out using a low cost workstation (DecStation 5000/240). The use of modular verification means that verifications can be performed interactively. All the experiments report a maximum of 6Mb of memory usage and are completed in a maximum of 30 seconds for the worst case.

3. Case study

The case study that we shall use is the "Steam Boiler Problem", that was suggested to the participants of the Dagstuhl Meeting on "Methods for Semantics and Specification" [**D95**].

The system is composed of a steam-boiler with four pumps, each having its own pump controller for supervision purposes, a device to measure the quantity of water in the steam-boiler, another for measuring the quantity of steam that comes out, and the corresponding operator desk and message transmission system.

The model of the system is not trivial, because the requirements specially emphasize the fault tolerant behavior. The system must continue working under the failure of some devices. In such a case, it must internally estimate the measures that cannot be read from the faulty devices. Also, the system must recognize internal failures such as measures outside the dynamic range of the steam-boiler, spontaneous changes of states, etc. A full explanation of the whole specification is beyond the goal of this paper. Please refer to [**D95**] for more details.

The following sections will describe the SA/RT model [1] (only the relevant parts that will be used for this verification case study: top level control, a pump and the water level measurement system).

3.1. Overview of the model. A first cut diagram is shown in the *Steam Boiler* chart (Figure 3) where we divide the system into processes. Circles drawned with dashed lines represent control transformations (that will be specified using state transition diagrams). The other circles represent either primitive data processes, or processes that are further decomposed in a lower level diagram. Arrowed dashed lines represent events (control flows) and continuous lines data flows. The *Steam Boiler* chart components are explained below (Figure 4):

- *Init Mode* is devoted to the initialization phase and will not be used later in this paper.
- *Control Modes* maintains the top level operation modes (*Init*, *Run* and *Emergency Stop*). In *Run* mode, the system can be in three different modes (*Normal*, *Degraded* or *Rescue*) depending on the particular combinations of failures of the devices.
- *Control Level* performs the main actions by opening and closing the pumps depending on the water level. The other processes control the behavior of each device (water and steam sensors, and pump), each one having its own diagram.

3.2. Pump. Figure 5 is the *Pump* chart, which is composed of the following control processes (Figure 6):

- *Pump Status* maintains the desired pump state that has been set from *Control Level*. *Pump Set Failure Status* maintains the failure/not failure status for the set of pump and pump controller
- Because pump and pump controllers have independent behavior and independent sources of failures, two processes (*Pump Failure Status* and *Pump Control Failure Status*) monitor the faulty states.
- Two more processes (*Pump Opening* and *Pump Closing*) are needed to model the manoeuvres of starting and stopping pumps, and internal failures detection.

---

[1] We have used a slight modification to the original SA/RT syntax for transitions, that consists in the possibility of representing conditions as composed of logical expressions on control flows and states of other processes.

For a process P, we denote the *Enable*, *Disable* and *Trigger* activators as $E.P$, $D.P$ and $T.P$ respectively. Also, $S.P$ is used to check the internal state of P.

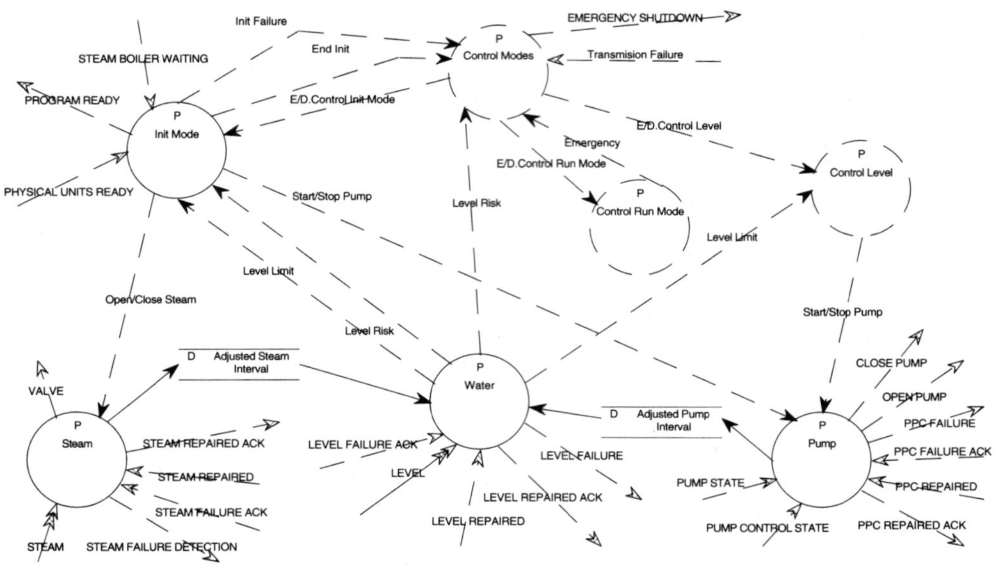

FIGURE 3. Steam Boiler

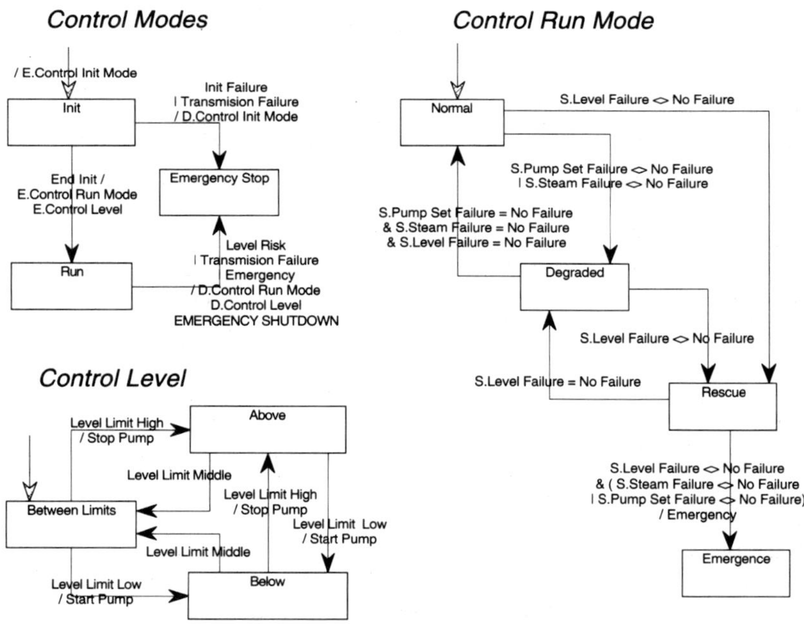

FIGURE 4. Steam Boiler Control

3.3. Water. *Water* is simpler than pumps. Its goal is to determine the quantity of wate r in the steam boiler and detect internal failures such as that the level measured is outside the dynamics of the system, or the absence of the *LEVEL* signal (see Figures 7 and 8).

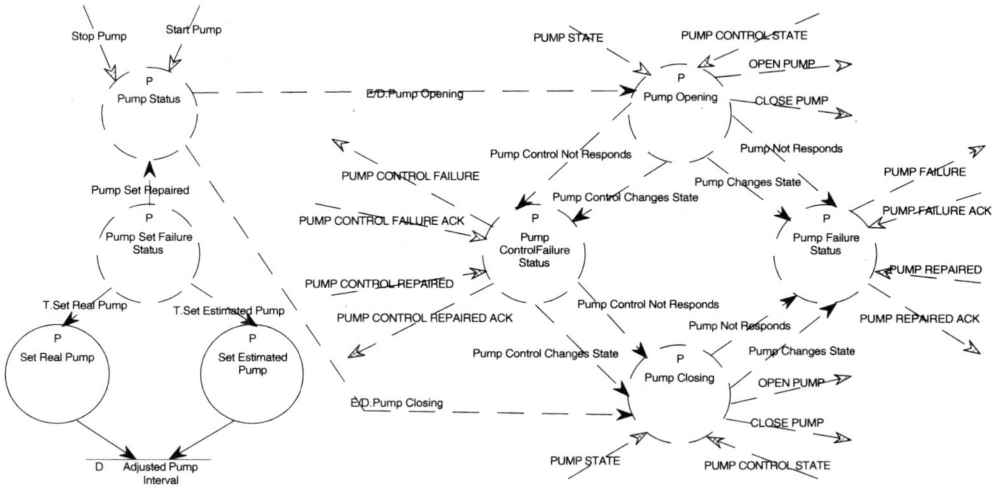

FIGURE 5. Pump

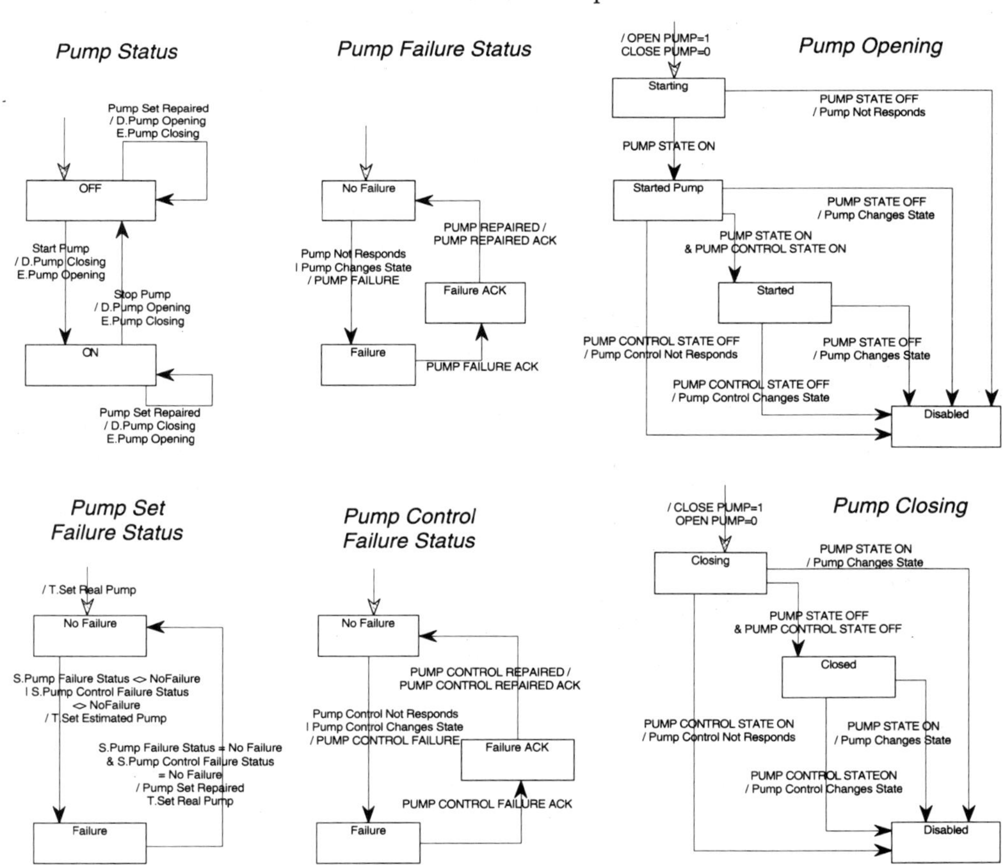

FIGURE 6. Pump Control

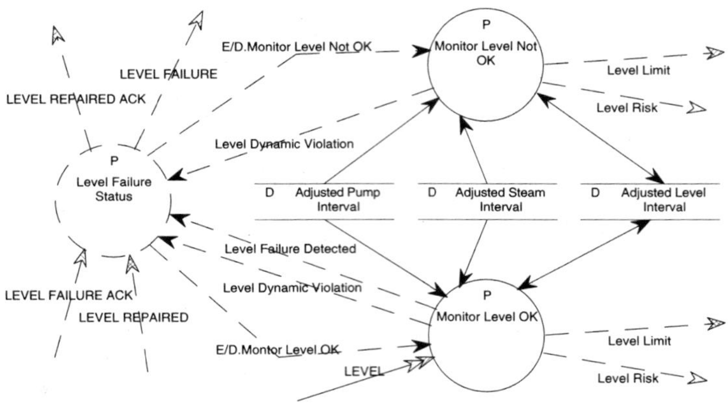

FIGURE 7. Water

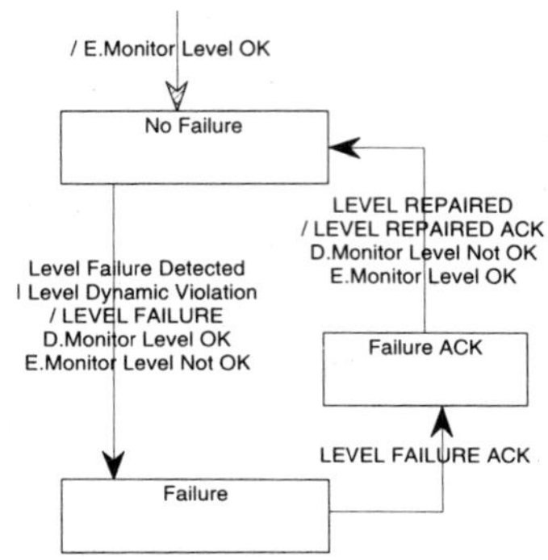

FIGURE 8. Level Failure Status

4. Verification

A large number of properties can be stated in order to safeguard the system described above. Part of them are explicitly included in the Dagstuhl specification, and part of them can be deduced from it. We shall concentrate our effort on a small set of representative properties. The verification process will be explained as if we were incrementally developing and interactively debugging the model using the verifier.

4.1. Properties to be verified. When the water level is low, the pump must be open. If not, we must detect a failure and the operating mode must be different to *Normal*. The exact mode (*Degraded*, *Rescue* or *Emergence*) depends on the combination of failures for different devices. Because the pump does not respond immediately, we consider the pump open if it is really open (*PUMP_STATE_ON*) or the open command (*OPEN_PUMP*) has been sent. The formula that must be verified is:

$$\phi_1 : \quad LEVEL = Low \rightarrow (OPEN\_PUMP \vee PUMP\_STATE\_ON \\ \vee ControlRunMode \neq Normal)$$

Also, we consider the symmetric situation when the level is high:

$$\phi_2 : \quad LEVEL = High \rightarrow (CLOSE\_PUMP \vee PUMP\_STATE\_OFF \\ \vee ControlRunMode \neq Normal)$$

Modular verification will be conducted by partitioning the system into three components as shown in the Table 1:

TABLE 1

Component	Processes in the structured model
MODES	*Control Modes* , *Control Run Mode*
WATER	*Water* (and its children), *Control Level*
PUMP	*Pump* (and its children)

Local formulas that we will prove are the following:

$$\phi_{P_1}: \quad PUMP \models INTERVAL(StartPump, StopPump) \rightarrow (OPEN\_PUMP \\ \vee PUMP\_STATUS\_ON \vee PumpSetFailureStatus \neq NoFailure)$$

$$\phi_{P_2}: \quad PUMP \models INTERVAL(StopPump, StartPump) \rightarrow (CLOSE\_PUMP \\ \vee PUMP\_STATUS\_OFF \vee PumpSetFailureStatus \neq NoFailure)$$

$$\phi_{L_1}: \quad WATER \models LEVEL = Low \rightarrow (INTERVAL(StartPump, StopPump) \\ \vee LevelFailureStatus \neq NoFailure)$$

$$\phi_{L_2}: \quad WATER \models LEVEL = High \rightarrow (INTERVAL(StopPump, StartPump) \\ \vee LevelFailureStatus \neq NoFailure)$$

$$\phi_M: \quad MODES \models (PumpSetFailureStatus \neq NoFailure \\ \vee LevelFailureStatus \neq NoFailure) \rightarrow ControlRunMode \neq Normal$$

It is clear that $\phi_{P_1} \wedge \phi_{L_1} \wedge \phi_M \rightarrow \phi_1$ and also $\phi_{P_2} \wedge \phi_{L_2} \wedge \phi_M \rightarrow \phi_2$. So, if we prove the above formulas we have proven our goals (ϕ_1 and ϕ_2). Because we observe that *WATER* and *PUMP* are closely related by means of the control flows *StartPump* and *StopPump*, a reasonable assumption for *PUMP* is that both events are never received simultaneously (in a micro-step). If we do not take this assumption into account now, we could be detect its need by examining the counter-examples produced in the *PUMP* verification. This assumption can be stated as $\neg(StartPump \wedge StopPump)$ or as two implications:

$$(StartPump \rightarrow StopPump = 0) \wedge (StopPump \rightarrow StartPump = 0)$$

This will be represented using the macro *ASSUME* as f ollows:

$$ASSUME(StartPump, StopPump, 0)$$

$$ASSUME(StopPump, StartPump, 0)$$

The set of properties for ϕ_1, assumptions and commitments can be represented graphically as in the Figure 9, which is self-explanatory:

4.2. Verification of PUMP. First, we execute the SPIN model checker for formula ϕ_{P_1}, including the above assumptions and the variables *StartPump* and *StopPump* as ghost variables.

The counter-example (trail) produced shows that *PUMP* receives *StopPump* and closes the pump. Then, it fails. The last macro-step enables us to detect the problem (its corresponding part of the trail is included below).

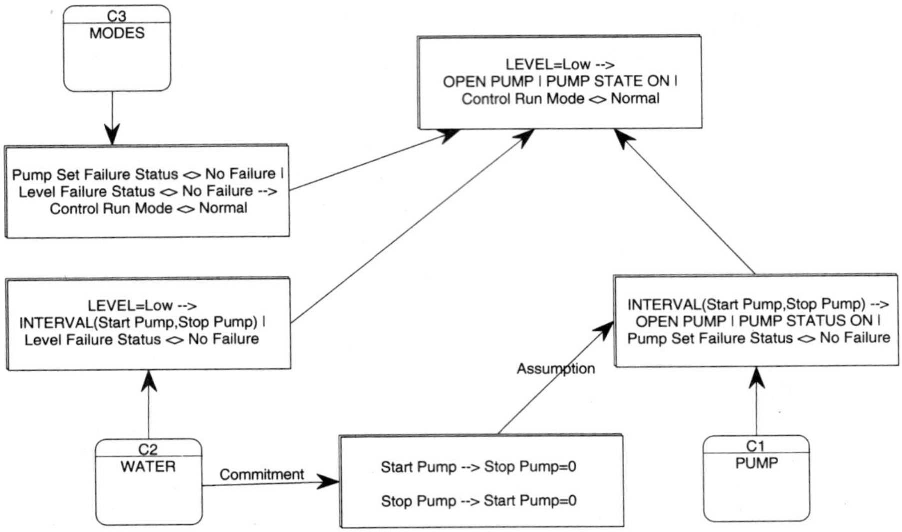

FIGURE 9. Graphical representation of properties for modular verification

```
*Macro Step: Environment
    **Received External Event PUMP_STATE_OFF
    **Received External Event PUMP_CONTROL_STATE_OFF
    **Received External Event PUMP_REPAIRED
  *Micro Step
    ** Transition T41: PumpFailureStatus: FailureACK->NoFailure,
       Action  PUMP_REPAIRED_ACK=1
  *Micro Step
    ** Ghost Event StartPump
    ** Transition T11: PumpSetFailureStatus: Failure->NoFailure,
       Action  PumpSetRepaired=1
  *Micro Step
    ** Transition T01: PumpStatus : OFF -> OFF,
       Action  PumpOpening=
Disabled;
       PumpClosing=Closing;
       CLOSE_PUMP=1 ; OPEN_PUMP=0
  *Micro Step
    **No Actions. End Macro Step
spin: line 2913 "paper.i", Error: assertion violated
spin: trail ends after 4371 steps
```

In this last macro-step the event *PUMP_REPAIRED* is received from the environment. The internal sequence of micro-steps that will set the pump to the normal state starts. But *StartPump* is sent as a consequence of some (ghost) action on the environment in the second micro-step. In the third step, the process *PumpStatus*, which is in the *OFF* state, simultaneously receives the commands *PumpSetRepaired* and *StartPump*. Due to the non-determinism, the system goes to the *OFF* state and tries to close the pump by sending *CLOSE_PUMP*. Assertion is violated.

If we remove the non-determinism by giving priority to the transition from *OFF* to *ON*, the problem will be solved. The solution is to change conditions in the transition from *OFF* to itself to $PumpSetRepaired \wedge \neg StartPump$. Because the closing of the pump is symmetric, the same problem would appear for ϕ_{P_2}, so we also change the transition from *ON* to itself to $PumpSetRepaired \wedge \neg StopPump$.

We run the verifier for the formula ϕ_{P_1} and it is shown to be true. Since the closing property of the pump is symmetric, and the graphical specification seems to be symmetric, a naive analyst might assume that the closing properties ϕ_{P_2} will also be true. But when we include them in the model checker we discover that they are false. The cause is that the problem is not truly symmetric because the *OFF* state in *PumpStatus* is the initial state and the *ON* state is not. The brief counter-example shows us that while in the initial state, *PumpClosing* is received, but *PUMP_STATE_ON* indicates that the pump is open and there is no transition that can be executed in response to *PumpClosing*. Note that this problem does not appear while *PumpStatus* is *ON*, because the system goes to this state in response to the first *StartPump*. A failure, when repaired, causes the *PumpSetRepaired* event to be received, and so, the pump will be open again.

We solve the problem by changing the transition from *OFF* to *OFF* to obtain the condition $(PumpSetRepaired \wedge \neg StartPump) \vee StopPump$. The transition from *ON* to *ON* does not need any Change. Additional properties which can be asserted and proven true of the desired response of the pump are:

$$WAS1(OPEN\_PUMP) \rightarrow \\ (PUMP\_STATE\_ON \vee PumpSetFailureStatus \neq NoFailure)$$

$$WAS1(CLOSE\_PUMP) \rightarrow \\ (PUMP\_STATE\_OFF \vee PumpSetFailureStatus \neq NoFailure)$$

or of the undesired res ponse:

$$INTERVAL(StartPump, StopPump) \rightarrow \\ (CLOSE\_PUMP = 0 \vee WAS0(StopPump))$$

$$INTERVAL(StopPump, StartPump) \rightarrow \\ (OPEN\_PUMP = 0 \vee WAS0(StartPump))$$

4.3. Verification of WATER. We follow the same procedure for the *WATER* component. In this case, *StartPump* and *StopPump* will not be ghost variables, and the assumptions will be proven using the corresp onding commitments[2]:

$$COMMIT(StartPump, StopPump, 0)$$

$$COMMIT(StopPump, StartPump, 0)$$

Formulas ϕ_{L_1} and ϕ_{L_2} are checked as true.

We can also state other interesting safety properties from the original specification. For instance, whenever *LEVEL* is detected to be outside the system dynamics, we consider the Level sensor to be failing, or if *LEVEL* is not present, the system detects a failure:

$$OutOfDynamics \rightarrow LevelFailureStatus \neq NoFailure$$

$$LEVEL = NotPresent \rightarrow LevelFailureStatus \neq NoFailure$$

[2]Recall that commitments are checked after each micro-step, as opposed to the system properties, which are checked at the end of each macro-step.

4.4. Verification of MODES. We verify the formula ϕ_M for *MODES* and find it true. Other properties that have been stated in the original specification are the invariant for the operating modes:

$$\begin{gathered}\mathit{ControlRunMode} = \mathit{Normal} \rightarrow \\ (\mathit{PumpSetFailureStatus} = \mathit{NoFailure} \wedge \mathit{SteamFailure} = \mathit{NoFailure} \\ \wedge \mathit{LevelFailure} = \mathit{NoFailure})\end{gathered}$$

$$\begin{gathered}\mathit{ControlRunMode} = \mathit{Degraded} \rightarrow \\ ((\mathit{PumpSetFailureStatus} \neq \mathit{NoFailure} \vee \mathit{SteamFailure} \neq \mathit{NoFailure}) \\ \wedge \mathit{LevelFailure} = \mathit{NoFailure})\end{gathered}$$

$$\begin{gathered}\mathit{ControlRunMode} = \mathit{Rescue} \rightarrow \\ (\neg(\mathit{PumpSetFailureStatus} \neq \mathit{NoFailure} \vee \mathit{SteamFailure} \neq \mathit{NoFailure}) \\ \wedge \mathit{LevelFailure} \neq \mathit{NoFailure})\end{gathered}$$

that also are found true.

Finally, we can conclude that our goal formulas ϕ_1 and ϕ_2 are true over the whole model.

4.5. Searching for non-progress cycles. One of the problems of the RSML semantics is the possibility of falling into infinite loops in which the system executes an infinite sequence of micro-steps. This type of potential livelock risk must be detected in order to safeguard performance. We must ensure that the system executes an infinite sequence of macro-steps. To do this, we have labeled the section of the PROMELA program in which the environment changes its value with a progress label, and by means of the SPIN option `-l` we can check the progress. We repeat the verification of each model using this option and we detect more problems:

SPIN detects that vvPUMP has a non-progress cycle, since *StartPump* and *StopPump* can appear infinitely often in the same micro-step (recall that they are ghost variables). Although this might not be a problem if *WATER* does not exhibit this behavior, it is a potential risk. The approach that we have taken is to restrict the behavior of ghost variables by stating that they appear at most, only once in the same macro-step:

$$\mathit{ASSUME}(\mathit{WAS0}(\mathit{StartPump}) \vee \mathit{WAS0}(\mathit{StopPump}) \rightarrow \mathit{StartPump} = 0)$$

$$\mathit{ASSUME}(\mathit{WAS0}(\mathit{StopPump}) \vee \mathit{WAS0}(\mathit{StartPump}) \rightarrow \mathit{StopPump} = 0)$$

On running the verifier with the `-l` option we detect the absence of non-progress cycles. Because these assumptions restrict the freedom of the *PUMP* behavior if we run the verifier without the `-l` option, we prove that properties are also true.

Now, using *WATER*, we try to discharge these new assumptions by translating them into commitments, and we find that *WATER* does not guarantee them. The counter-example shows us that *LevelFailure* is in *Failure* state. *MonitorLevelNotOK* data process can determine (in the first micro-step) the level limits that are received by *ControlLevel* which sends a Start/Stop command to the pump. But if *LEVEL_REPAIRED* is received, Data Transformation *MonitorLevelOK* is enabled and another start/stop message is sent (in the second micro-step). We must transform the assumption/commitment into another "softer" one:

$$\mathit{ASSUME}(\mathit{WAS0}(\mathit{StartPump}) \rightarrow \mathit{StartPump} = 0)$$

$$\mathit{ASSUME}(\mathit{WAS0}(\mathit{StopPump}) \rightarrow \mathit{StopPump} = 0)$$

We verify the above formulas both for *PUMP* as assumptions and for *WATER* as commitments, and re-check the existence of non-progress cycles. Similar assumptions must be made for the *MODES* component by assuming that in the same micro-step, transitions from/to the *NoFailure* states are never taken more than once, along with the corresponding commitments for *PUMP* and *WATER*. Now the specification is found to be true.

5. Conclusions

We have shown how the SPIN model checker can be used in conjunction with graphical specification methods, and how we can use it for debugging the specification as well as proving some interesting properties in said specification. The modular approach allows us to perform separate verifications for components of the model and compose them in order to prove properties for the whole model. The PROMELA program that represents the model is slightly more complicated and less straightforward than if it were coded directly from the textual specification as in [**DC95**], but its great advantage is that it is automatically obtained from a graphical specification, and so, easier to develop, review and maintain.

Our present work, as part of the AUTOVER project and other ongoing projects, is addressed towards a complete integration of structured development tools and model checkers. The user specifies the system graphically and the results of the verification (counter-examples) can be graphically animated in order to debug the model. We also represent the whole set of properties to verify in a similar way to that shown in Figure 9. This allows us to perform regression verifications and detect the problems caused by changes in the model. We also plan to compare SPIN with other model checkers [**C96**], and to study the possibilities of automatic reductions of the model [**SR94**] and relation checking [**E95**], both for the part of model that is being verified as well as for its environment.

References

[AL93] M. Abadi, L. Lamport, *Conjoining Specifications.* DEC Software Research Centre. Research Report no. 118, 1993.

[A95] A. Alonso, L. Baresi, H. Christensen, M. Heikkinen, *IDERS: An Integrated Environment for the Development of Hard Real-Time Systems.* Euromicro Workshop on Real-Time Systems. Odense, Denmark, 1995.

[C96] J, C. Corbett, *Evaluating Deadlock Detection Methods for Concurrent Software.* IEEE Transactions on Software Engineering, Vol. 22, no. 3, pp. 161-180, 1996.

[CGR93] D. Craigen, S. Gerhart, T. Ralston, *An International Survey of Industrial Applications of Formal Methods.* U.S. Department of Commerce, Technology Administation. National Institute of Standards and Technology, Computer Systems Laboratory Gaithersburg. Report NISTGCR 93/626, 1993.

[D94] W. Damm, H. Hungar, P. Kelb, R Schlör, *Using Graphical Specification Languages and Symbolic Model Checking in the Verification of a Production Cell.* In: Case Study "Production Cell". Claus Lewerentz and Thomas Lindner Eds., pp. 89-107, 1994.

[D95] *Seminar on "Methods for Semantics and Specification"* Dagstuhl, Wadern, Germany, on June 5-9, 1995. Available in http://www.informatik.uni-kiel.de/~ procos/dag9523/dag9523.html

[D93] N. Day, *A Model Checker for Statecharts (Linking CASE Tools with Formal Methods).* MS Thesis, University of British Columbia, Dept. of Computer Science (Tech. Report 93-35), 1993.

[DR96] D. Dill, J. Rushby, *Acceptance of Formal Methods: Lessons from Hardware Design.* IEEE Computer, vol. 29, no 4, pp. 23-24, 1996.

[DC95] G. Duval, T. Cattel, *Specifying and Verifying the Steam Boiler problem with Spin.* In [**D95**], 1995.

[FMM94] M. Felder, D. Mandrioli, A. Morzenti, *Proving Properties of Real-Time Systems Through Logical Specifications and Petri Net Models.* IEEE Transactions on Software engineering, Vol. 20, no. 2, pp. 127-141, 1994.

[Ha87a] D. Harel, *STATECHARTS: A Visual Formalism for Complex Systems.* Science of Computer Programming, North Holland, Vol. 8, pp. 231-274, 1987.

[Ha87b] D. Harel, A. Pnuelli, J. Pruzan-Schmidt, R. Sherman, *On the formal semantics of Statecharts.* Proceedings of the 2nd Symposium on Logic in Computer Science, pp. 56-64, 1987.

[Ha90] D. Harel, H. Lachover, A. Naamad, A. Pnueli, M. Politi, R. Sherman, M. Tachtenbrot, *Statemate: a working environment for the development of complex reactive systems.* IEEE Transactions on Software Engineering, Vol. 16, no. 4, pp. 403-414, 1990.

[HP87] D. J. Hatley, I. Pirbhai, *Strategies for Real Time System Specification.* Dover Press, New York, 1987.

[E95] H. Erdogmus, *Verifying Semantic Relations in* SPIN. Proceedings of the 1st SPIN Workshop. Montreal, 1995.

[Ho91] G. J. Holzmann, *Design and Validation of Computer Protocols.* Prentice-Hall, 1991.

[Hu91] G. Huizing, *Semantics of Reactive Systems: Comparison and Full Abstraction.* Ph.D. Thesis, Technische Universiteit of Eindoven, 1991.

[L93] G. León, J. C. Dueñas, J.A. de la Puente, A. Alonso, *The IPTES Environment: Support for Incremental, Heterogeneus and Distributed Prototyping.* Real-Time Systems Journal, pp. 153-172, 1993.

[LL95] S. Leue, P. B. Ladkin, *Implementing Message Sequence Charts in* PROMELA. Proceedings of the 1st SPIN Workshop. Montreal, 1995.

[LL96] S. Leue, P. B. Ladkin, *Implementing and Verifying Scenario-Based Specifications Using* PROMELA/XSPIN. Proceedings of the 2nd SPIN Workshop. New Brunswick, NJ, 1996.

[L94] N. G. Leveson, M. P. E. Heimdahl, H. Hildreth, J. D. Reese, *Requirements Specification for Process Control Systems* IEEE Transactions on Software Engineering, Vol. 20, no. 9, pp. 684-707, 1994.

[SR94] R. Simone, A. Ressouche, *Compositional Semantics of Esterel and Verification by Compositional Reductio ns.* In: Computer Aided Verification, Springer LNCS-818, 1994.

[T94] J. Tuya, *Specification and Verification of Reactive Systems Using Structured Methods and Temporal Logic.* Ph.D. Thesis. Universidad de Oviedo, 1994.

[T95] J. Tuya, L. Sánchez, J. A. Corrales, *Using a Symbolic Model Checker for Verify Safety Properties in SA/RT Models.* 5th European Software Engineering Conference. Springer LNCS 989, pp. 59-75, 1995.

[WM85] P. T. Ward, S. J. Mellor, *Structured Development for Real-Time Systems.* Prentice-Hall, 1985.

[W86] P. T. Ward, *The Transformation Schema: An Extension of the Data Flow Diagram to Represent Control and Timing.* IEEE Transactions on Software Engineering, Vol. 12, no. 2, pp. 198-210, 1986.

Universidad de Oviedo, Campus de Viesques, E.T.S. de Ingenieros Industriales e Informáticos, Carretera de Castiello, s/n. E-33394 GIJON / SPAIN
E-mail address: tuya@etsiig.uniovi.es

Universidad de Oviedo, Campus de Viesques, E.T.S. de Ingenieros Industriales e Informáticos, Carretera de Castiello, s/n. E-33394 GIJON / SPAIN
E-mail address: dediego@etsiig.uniovi.es

Universidad de Oviedo, Campus de Viesques, E.T.S. de Ingenieros Industriales e Informáticos, Carretera de Castiello, s/n. E-33394 GIJON / SPAIN
E-mail address: claudio@etsiig.uniovi.es

Universidad de Oviedo, Campus de Viesques, E.T.S. de Ingenieros Industriales e Informáticos, Carretera de Castiello, s/n. E-33394 GIJON / SPAIN
E-mail address: corrales@etsiig.uniovi.es

DIMACS Series in Discrete Mathematics
and Theoretical Computer Science
Volume **32**, 1997

Memory Efficient State Storage in SPIN

Willem Visser and Howard Barringer

ABSTRACT. The use of an Ordered Binary Decision Diagram (OBDD) to store all visited states during on-the-fly model checking (or reachability analysis) is investigated. To improve the time and space efficiency a state compression technique is introduced. This compression technique is safe, in the sense that no two unique states will have the same compressed representation. A number of examples are used to evaluate an experimental implementation of the OBDD state store within the SPIN validation tool. In all the examples a reduction in space is achieved when using the OBDD state store as opposed to the more traditional hash table state store. The memory and time usage when combining partial orders with the OBDD state store is also considered.

1. Introduction

Temporal logics can express changes over time without introducing time explicitly and are therefore suitable for specifying many correctness properties of concurrent systems. Since many interesting programs can be modelled as finite-state systems, it was a significant development when algorithmic methods were discovered to verify temporal properties of finite-state systems [**CES83**]. Since that time a variety of concurrent systems have been modelled and automatically verified via such methods, these include: communication protocols, hardware designs and operating system kernels. A finite labelled state-transition graph (also called a *Kripke* structure) is used to represent the behaviour of concurrent system and propositional temporal logic formulae are used to express desired properties of the system. A so-called *model checker* is then used to check whether the Kripke structure satisfies (is a model of) the temporal formula specifying the required behaviour. Surveys of model checking techniques can be found in [**Wol87**] and [**CGL93**].

1991 *Mathematics Subject Classification.* Primary 68Q60; Secondary 68N99, 68U30.

Key words and phrases. On-the-fly Reachability Analysis, State Space Compression, Binary Decision Diagrams, SPIN.

The first author is supported by the FRD (South Africa) and the Department of Computer Science, University of Manchester.

Early model checking algorithms [**CES86**] required the complete state graph to be generated before-hand and kept in memory throughout the model checking process. Because of the so-called *state explosion* problem (the number of reachable states grow exponentially in the number of concurrent components) when analysing sufficiently large concurrent systems, the state graph is however often too large to fit in memory. This limits the size of the systems that can be model checked with these algorithms. One possible solution is to generate only the part of the state graph required to validate (or invalidate) the given temporal logic requirement. This technique is commonly referred to as *on-the-fly* model checking, since the state graph is generated during model checking [**VW86, BVW94**]. On-the-fly algorithms generate the state space in a depth-first manner and keep track of all reached states to avoid doing unnecessary work. The boundaries of on-the-fly techniques were advanced considerably when it was realised that much of the state explosion problem was due to the generation of all interleavings of independent transitions in different concurrent components. *Partial order reduction* techniques were therefore introduced to ensure that many of these unnecessary interleavings are not explored during state generation [**Val90, GW93, Pel94, HP94**].

Perhaps the greatest advancement in reducing the limitations imposed by the state explosion problem was made with the advent of *symbolic model checking* [**BCM$^+$90, McM92**]. In symbolic model checking the transition relation is no longer represented explicitly, but rather implicitly by encoding it with *ordered binary decision diagrams* (OBDDs) [**Bry86**]. The temporal logic formula to be checked is translated into its fixed point representation. The model checking algorithm entails calculating these fixed points by performing operations on the OBDD that represents the transition relation. After each iteration of the fixed point calculation the set of states obtained are again represented by an OBDD. The main advantage of this algorithm is therefore that the state graph is never explicitly constructed. If the state space of a system exhibits some form of regularity then the OBDD representation of a set of its states can be very compact. Symbolic model checkers are not goal directed, since the complete state space is generated during model checking.

Unfortunately, both on-the-fly techniques and those using OBDD encodings have potential drawbacks. The main problem with on-the-fly algorithms is the storage space required to record already visited states during state generation. For OBDDs, on the other hand, the size of an OBDD is heavily dependent on the ordering of the boolean variables within it. An even bigger disadvantage of OBDDs is that certain boolean functions will always have an exponentially sized OBDD regardless of the variable ordering used [**Bry92**]. Integer multiplication is an example of such a function. The size of the OBDD representing the transition relation can therefore cause the symbolic model checking to become impractical.

In this paper we explore a combination of on-the-fly model checking and OBDD encodings. The on-the-fly ethos will be adopted: the algorithm will be goal directed and therefore the state space will be generated during the model checking process. The transition relation will therefore not be represented by an OBDD, but, one will be used to represent the set of states already visited during the state generation. If there is any regularity in the state space generated this would be reflected in the size of the OBDD, i.e. its size will increase polynomially, rather than exponentially, with the increase in states visited. OBDD based model checkers have until now been used mostly for hardware verification, since hardware systems usually exhibit regular state spaces. These systems also tended to be synchronous. Since asynchronous software (and hardware) systems have not received the same amount of attention from the symbolic model checking community, it was decided to apply our technique to these type of systems. The effect of using partial order techniques in combination with OBDDs will also be investigated.

The outline of the paper is as follows. Section 2 introduces the basic on-the-fly algorithm used during state generation. The next two sections are devoted to explaining our use of OBDDs (section 3) and how it is combined with the SPIN [**Hol91**] on-the-fly validation system (section 4). In section 5 a novel way of state compression is introduced to allow more efficient OBDD representations. In sections 5.2 and 6, the results achieved by the technique are discussed as well as the effect of partial order reductions on the OBDD size. The last section contains conclusions and some future extensions to the algorithm.

2. On-the-fly Model Checking

Although it has long been known that linear time temporal logic can be efficiently model checked by only generating relevant parts of a state graph [**VW86**], it has only recently been shown that the same is true of the branching time temporal logics CTL and CTL* [**BVW94, BCG95**]. Here we will only concentrate on efficient state storage during on-the-fly model checking and therefore not be concerned about the temporal logic being used.

In Figure 1 the standard depth-first search algorithm is shown that implements the state generation during model checking. When used during model checking the iterative form of the algorithm is used and after each state, s, is generated the model checker will determine the truth-value of the current temporal logic subformula from the value assignments within s. Since the focus here is on state generation, rather than on the model checking algorithm, and since recursive algorithm in Figure 1 is easier to follow than the iterative version we will use the recursive one to explain the basic issues involved. Two data structures are used within the search: a set of all states visited during the search, *VisitedStates*, and a stack of states to keep track of the current execution path, *Stack*. During initialisation the initial state, s_0 is entered into *VisitedStates* as well as pushed onto *Stack*. The depth-first search is then invoked by calling *dfs()*. In *dfs* all transitions enabled in the current state, s, at the top of the stack are executed in a depth-first manner. If the first successor state s' of s has not been reached before, i.e. it is not in *VisitedStates*, it is entered into *VisitedStates* and pushed onto *Stack*. By calling *dfs* again s' now becomes the current state. When all the successor states from the current state s have been visited then s is removed from the stack and the previous state on the stack becomes the current state.

The representation of the set *VisitedStates* has been the focus of much research [**Hol88, GHP92, WL93, Gre96**]. This is not surprising since this is the part of on-the-fly model checking that will determine its tractability when checking large designs. The most commonly used method is to represent the set as a large hash

```
PROCEDURE DFS()

   PROCEDURE dfs()
   BEGIN
      s = top(Stack);
      FOR all transitions t enabled in s DO
          s' = successor(s) after executing t;
          IF s' NOT IN VisitedStates THEN
             Enter s' into VisitedStates;
             Push s' onto Stack;
             dfs()
          END
      END;
      Pop s from Stack
   END dfs;

BEGIN
   Enter s0 into VisitedStates;
   Push s0 onto Stack;
   dfs()
END DFS;
```

FIGURE 1. Depth-first State Generation during Model Checking.

table of states. When a new state is generated it is hashed to obtain the index into the table. Since the states must be stored in its entirety, to allow for comparisons during the resolving of possible hash conflicts, this method is not very efficient when a large number of states must be stored. The most novel approach to date is the so-called *bitstate hashing* [**Hol88**] used in the SPIN validation system. This technique requires a large vector of bits (of fixed size) to be maintained in memory to keep track of previously generated states. A hashing technique is used to compute an index into this bit vector from the value of each state. However, since the validation results can be invalid when a hash conflict occurs this technique is not always desirable when validating safety critical systems. The coverage obtained with bitstate hashing can be improved by either increasing the number of hashing functions used or using more than one bit to hash into (*hashcompact* [**WL93**]). A summary of bitstate hashing as well as comparisons with the hashcompact method can be found in [**Hol95**].

In the next section we introduce OBDDs as a possible alternative to the above mentioned representations of *VisitedStates*. After implementing the OBDD to record visited states in the SPIN validation system a further state compression method was added to reduce both the time and memory requirements of this method. This compression method can also be used with the traditional hash table methods to reduce the memory required, at the price of adding a small constant factor to the run-time.

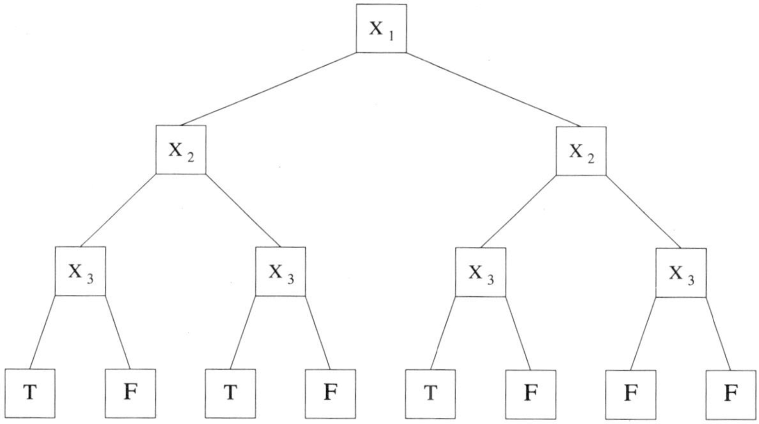

FIGURE 2. Decision tree for the boolean function $f(x_1, x_2, x_3) = \overline{x}_1 \cdot \overline{x}_2 \cdot \overline{x}_3 + \overline{x}_1 \cdot x_2 \cdot \overline{x}_3 + x_1 \cdot \overline{x}_2 \cdot \overline{x}_3$

3. OBDD State Store

OBDDs are directed acyclic graphs that represent boolean functions in a canonical form. As an example, let us consider the OBDD that represents the function $f(x_1, x_2, x_3) = \overline{x}_1 \cdot \overline{x}_2 \cdot \overline{x}_3 + \overline{x}_1 \cdot x_2 \cdot \overline{x}_3 + x_1 \cdot \overline{x}_2 \cdot \overline{x}_3$, where $\cdot$ denotes the AND operation and $+$ the OR operation. The decision tree for this function is given in figure 2. Left branches from a node indicates the variable is 0 and similarly right branches indicate value 1. Terminal nodes are labelled with T for true and F for false. To construct the OBDD representation of f a total ordering on the variables of f must be imposed. In figure 2 this ordering is $x_1 < x_2 < x_3$. There are three transformation rules on these graphs, that do not alter the function represented, but may reduce their size. An OBDD is the name given to the graph that cannot be reduced any further. The rules are:

Remove Duplicate Terminals: There must remain only one terminal with a given label. All arcs to duplicate labelled terminals must be redirected to the remaining one. A further optimisation is not to show any arcs leading to F, although they are implicitly there. In figure 3 this optimisation is performed on the decision tree for f.

Remove Duplicate Nonterminals: If two nonterminal nodes have the same label and their left and right branches are the same then one must be removed. All arcs must be redirected to the remaining one. This is shown in figure 4 where the three nodes labelled x_3 (with left arc to T and right to F) in figure 3 have been replaced by one x_3 node.

Remove Redundant Tests: If both arcs from a nonterminal, say n, point to the same node, say n', then n must be removed and all its incoming arcs must be redirected to n'. In figure 4 the left-hand x_2 is such a redundant node since both its left and right arcs point to x_3. The OBDD for f is shown in figure 5 after this last transformation is performed.

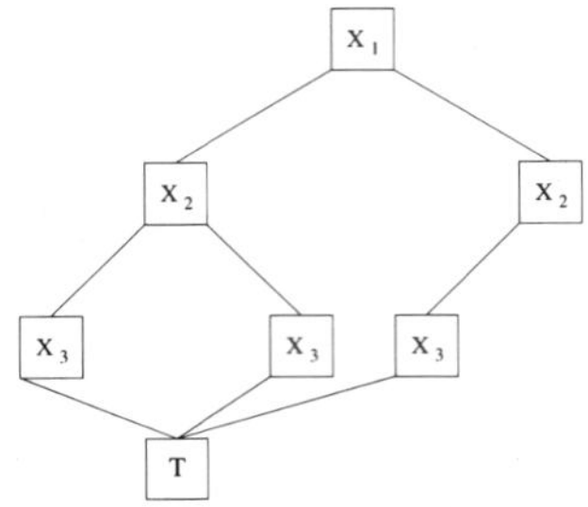

FIGURE 3. Removing duplicate T terminals and explicit F terminals from the decision tree of f

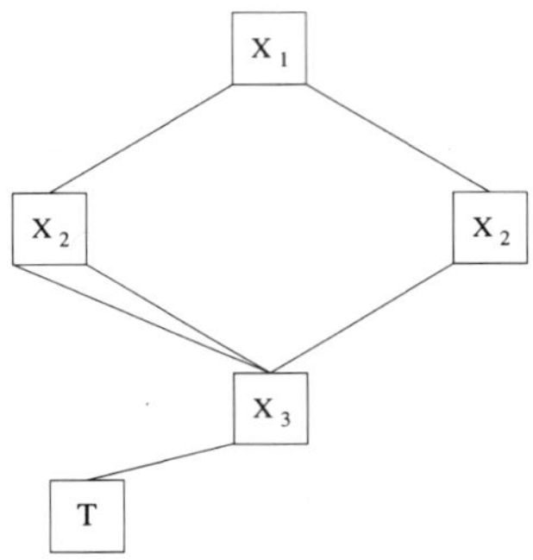

FIGURE 4. Removing duplicate nonterminals.

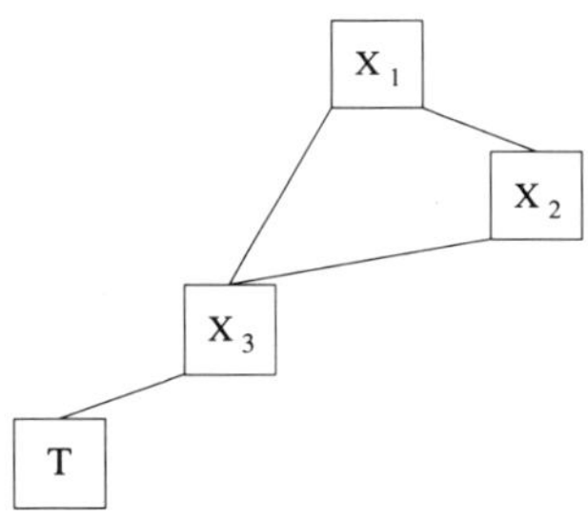

FIGURE 5. OBDD for function f after removing redundant tests.

These rules must be applied repeatedly, since the application of one rule can cause more transformations to be possible on the resulting graph. It will now be shown how an OBDD can be used to represent the set of states visited during on-the-fly model checking. First, recall that the state, s, of a system consists of a vector of bits, $s = x_0, x_1, \cdots, x_n$, where x_i is bit i in this so-called state vector. If a state, s, is visited during a traversal it can be represented by the boolean function $f(s) = 1$, or, in other words, $f(x_0, x_1, \ldots, x_n) = 1$. For example, consider a state vector with only three bits, x_1, x_2 and x_3 and a state is written as $< x_1x_2x_3 >$. Assume the following states are visited $< 000 >$, $< 010 >$ and $< 100 >$. Then the OBDD for the boolean function $f(x_1, x_2, x_3) = \overline{x}_1 \cdot \overline{x}_2 \cdot \overline{x}_3 + \overline{x}_1 \cdot x_2 \cdot \overline{x}_3 + x_1 \cdot \overline{x}_2 \cdot \overline{x}_3$ will represent the states visited. Thus if the variable ordering $x_1 < x_2 < x_3$ is assumed then the

OBDD in figure 5 represents the visited states for this example. To add a state, s, to the set of visited states it is therefore necessary to first convert s to its OBDD representation and then to perform an OR-operation on this OBDD and the one representing the set of states already visited. The resulting OBDD will represent the set of states with s added. To establish if a new state, s, is in the set of already visited states, the OBDD representing the visited states is traversed according to the values of the bits in s. For example, to check if the state $s =< 010 >$ is in the OBDD of figure 5 we check if node x_1 has a left arc–since $x_1 = 0$ in s–this arc is traversed to reach node x_3; now since $x_3 = 0$ in s the left arc of x_3 is traversed and the resulting terminal node T is found indicating s is in the OBDD. If the existence of state $s =< 011 >$ was checked then at node x_3 the right branch would have been taken leading to the terminal F indicating s has not been visited yet. Only the following three operations are required to implement an OBDD state storage mechanism during on-the-fly model checking:

EncodeOBDD(s): This function takes as input a state and returns the OBDD encoding of the state.

OR(OBDD1,OBDD2): Returns the OBDD that is obtained after applying the transformation rules to the result of adding OBDD1 to OBDD2.

Exists(OBDD1,s): Returns true if state s is in OBDD1 else false.

The advantage of using an OBDD to record visited states becomes apparent when large parts, or even, the complete reachable state space of a model is visited. Since this will in most cases cause the OBDD to become smaller, due to the applying of the transformation rules. For example consider the previous example where the states $< 000 >$, $< 010 >$ and $< 100 >$ were visited. If we now assume the rest of the reachable states of this example (the five states $< 001 >$,$< 011 >$, $< 101 >$,$< 110 >$ and $< 111 >$) are visited as well then the OBDD representing the visited states will only contain the terminal node T.

3.1. Example. Consider the mutual exclusion problem for two processes where each process ($i = 1, 2$) can be in one of three code regions: noncritical (N_i), trying (T_i) or critical (C_i). A binary semaphore S is used to protect the critical region. The value of the semaphore is indicated by S_i, where i can be 0 or 1. A process can only enter its *critical* region from its *trying* region if the value of the semaphore is 0. When a process enters its critical region the value of the semaphore becomes 1 and on leaving the critical region and entering the *noncritical* region the value of the semaphore becomes 0 again. The state of the system consists of the values of the control variable for $process_1$, $process_2$ and the *semaphore* variable and is described by $(process_1, process_2, semaphore)$. An asterisk ("*") in the guard of a transition indicates that the field can hold any value for the guard to be satisfied in that state. The action part of the transition will indicate how each field is changed, with "*" indicating unchanged fields. The start state is (N_1, N_2, S_0) and the transitions are the following:

$$
\begin{array}{lrcl}
(1) & (N_1, *, *) & \rightarrow & (T_1, *, *) \\
(2) & (T_1, *, S_0) & \rightarrow & (C_1, *, S_1) \\
(3) & (C_1, *, *) & \rightarrow & (N_1, *, S_0) \\
(4) & (*, N_2, *) & \rightarrow & (*, T_2, *) \\
(5) & (*, T_2, S_0) & \rightarrow & (*, C_2, S_1) \\
(6) & (*, C_2, *) & \rightarrow & (*, N_2, S_0)
\end{array}
$$

The two control variables each have three possible values: N_i (0), T_i (1) or C_i (2). The semaphore has two values: S_0 (0) or S_1 (1). The two control variables will thus have two bits and the semaphore one bit allocated for it in the state vector (Figure 6).

State vector: 5 bits		
$Process_1$	$Process_2$	Semaphore
4..3	2..1	0

FIGURE 6. A state vector for the mutual exclusion system

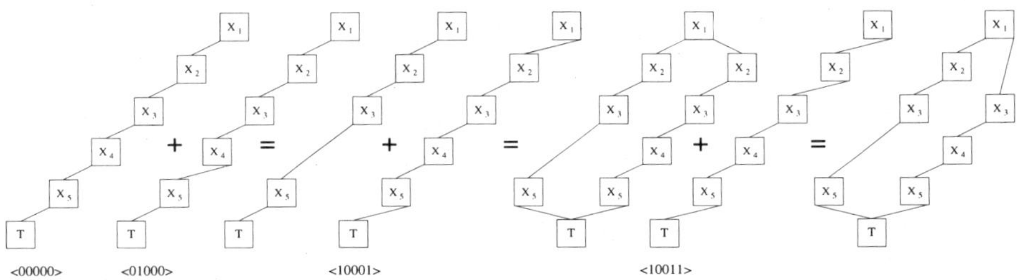

FIGURE 7. OBDD representations for adding the first four states of the mutual exclusion system to the state store.

During initialisation the initial state (N_1, N_2, S_0), which corresponds to the state vector $< 00000 >$, is inserted into the set of visited states. The next state to be visited is (T_1, N_2, S_0) ($< 01000 >$) which is generated after executing transition 1. The OBDD resulting from adding this state to the initial state is shown on the left-hand of figure 7. In the rest of figure 7 the resulting OBDDs are shown after adding the next three states visited in the depth-first search. The OBDD on the right therefore contains the states $< 00000 >$, $< 01000 >$, $< 10001 >$ and $< 00010 >$. In figure 8 the OBDD is shown after each of the remaining four states of this model is added to the set of visited states.

The memory requirement for this method is that of the largest OBDD generated during the search. For this example it is the memory required to represent the second and third last OBDDs (see figure 8) which both have 12 nodes (excluding the implicit F-terminal). Considering that the largest OBDD for a boolean function with five variables is 14, this example performs close to the worst-case[1]. In general

[1]If n is the number of variables in a boolean function then there are $3 * 2^{\frac{n}{2}} - 2$ nodes when n is even and $2 * (2^{\lceil \frac{n}{2} \rceil} - 1)$ nodes if n is odd in the largest possible OBDD for the function (excluding the implicit F-terminal).

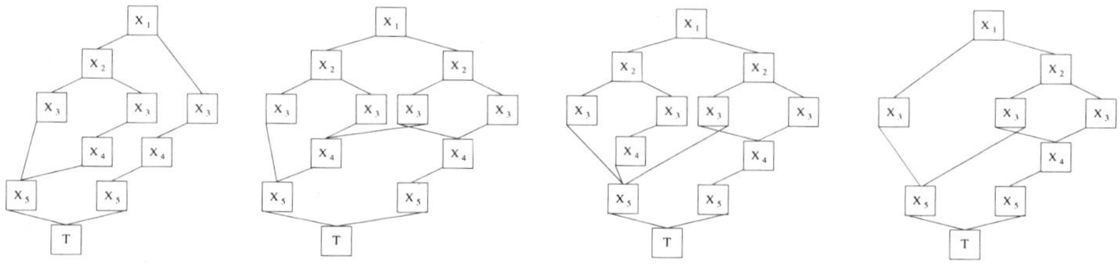

FIGURE 8. OBDD representations after the last four states of the example are added. The rightmost one represents all eight reachable states of the system.

the OBDD representing the state store will have a best-case performance that is *linear* and a worst-case that is *exponential* in the number of bits in the state vector. As it is unclear what the heuristics should be for choosing a variable ordering, we will only consider one variable ordering in the sequel. The variable order will be the same as the order of the bits in the state vector.

4. SPIN with OBDD State Store

With the aid of the BDD package[2] from Carnegie Mellon the two functions EncodeOBDD and Exists were implemented and the `bdd_or` function provided were used to add two OBDDs together. The functions were added to the SPIN[3] tool in a straight-forward fashion to obtain an experimental implementation for an on-the-fly model checker that uses an OBDD to record visited states.

The first example on which the method is evaluated is a model consisting of n concurrent counters, where each counter loops through the values 0 to 9. This example was chosen because it exhibits a regular state space and would therefore allow the OBDD state store to perform close to optimal. In Table 1 the results for 4, 5 and 6 concurrent counters are shown when using the OBDD state store compared with the table storage. The rest of the examples in Table 1 are all real-world models: 'Address Interface' is a model of part of an asynchronous microprocessor [**FDG$^+$93**]; 'snoopy' is a cache coherence model available with the SPIN system; 'scheduler' is a model of a process scheduler that forms part of a commercially available micro-kernel [**FdV91**]; 'pftp' is a protocol model also available with the SPIN system. The comparison is done on the memory used in megabytes and the time[4] taken in seconds for the two methods. The number of nodes in the largest OBDD required, the number of bits in the state vector and the number of reachable states are also indicated. As expected the counter examples show the biggest improvement when using the OBDD state store. The memory for the table store increase exponentially (since the number of states increase exponentially) while the OBDD only require a linear increase, in the number of concurrent counters.

---

[2]Available by ftp from emc.cs.cmu.edu (pub/bdd/bddlib.tar.Z)

[3]Available by ftp from netlib.att.com (netlib/spin/spin2.8.Src.tar.Z)

[4]The tests were run on a SPARC10 with 64Mb of memory.

Model	Table Store		OBDD Store			Bits	States
	Mem (Mb)	Time (s)	Mem	Nodes	Time		
counter 4	0.1	11	0.3	158	358	72	10^4+1
counter 5	1	647	0.32	233	1580	88	10^5+1
counter 6	32	3012	0.33	283	18780	94	10^6+1
addr interface	0.7	2.8	0.74	22470	580	608	9491
snoopy	7.7	100	9.97	421165	36824	704	91920
scheduler	14.7	202	0.95	53354	36259	472	256929
pftp	48	600	>36	>10^6	>60000	928	439895

TABLE 1. Comparison between an OBDD and a (hash) table for storing visited states in SPIN.

Unfortunately, for the real-world examples the OBDD state store does not perform as well as in the contrived examples of concurrent counters. For example the generation of the state space for the pftp model was aborted after 16 hours, since only half the number of states had been generated in that time. The reason for this is that the performance of the OBDD store is critically dependent on the number of bits in the state vector. Since every state generated during the search must first be checked against the visited states in the OBDD and then encoded in an OBDD if it has not been visited, the Exists and EncodeOBDD functions are dependent on the state vector size. Therefore, since the size of the OBDD representing a new state is dependent on the number of bits in the state vector, so too will the `bdd_or` function be dependent on the state vector length. In the next section a compaction scheme is presented that compresses the length of the state vector, thereby allowing the OBDD state store to be more effective.

5. State Compression

A model that has n bits in its state vector, seldom visits all 2^n states. In SPIN this is principally because of the fact that only predefined types are available. Therefore when a variable will only be assigned values in the range $0 \ldots 7$, it would still be required to be of type *byte* which would mean 8 bits in the state vector instead of the required 3. In general, however, it is because variables in different concurrent processes do not "interleave" all their possible values. Finding ways of reducing the state vector size, without losing information, would therefore seem a worthwhile pursuit. Strangely, however, *probabilistic* methods [**Hol95, WL93, SD95**] are attracting more attention than *safe* reduction methods [**Vis93**] in the research community. Probabilistic methods allow the state vector to be compressed in such a fashion that different states might have the same compressed state. During on-the-fly model checking this might cause a search to be truncated prematurely in which case certain parts of the state space might be ignored. Here a safe reduction method will be introduced that will reduce the memory requirements for a large class of models at the cost of an acceptable run-time overhead.

The intuition behind the method is first described. In the same fashion that a model will seldom visit all its 2^n possible states, so too will a process seldom visit all of its potential states. Furthermore, a state explosion is seldom caused by

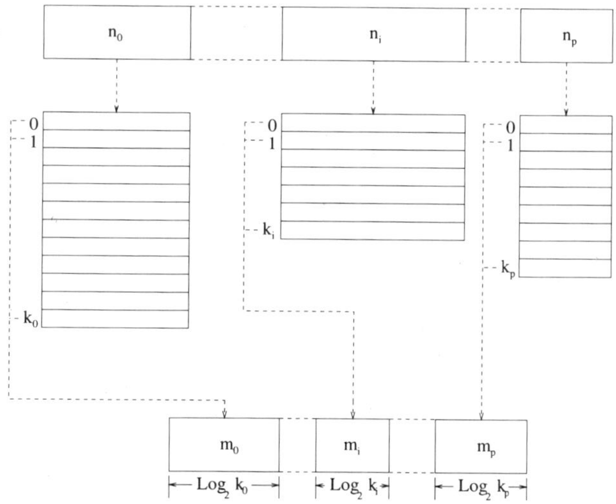

FIGURE 9. State vector compression via intermediate tables.

one process visiting a large number of states, it is more commonly caused by the cartesian product of a number of "small" concurrent processes. Now if we consider a process that is allocated m bits in its state vector but visits only $k = 2^n$ states where $k \ll 2^m$ then $m-n$ bits will be wasted every time a state is stored during the validation. This becomes more significant when this process is part of a model that generates a large number of states, since $m - n$ bits will be wasted not only when the process generates a new state, but whenever any of the other processes in the model cause a new state to be generated. It would therefore seem sensible to store only each of the 2^m states for the process once and only refer to them indirectly when the states are stored. This is achieved by storing the k m-bit values in a table and using the index in the table as the state in the state vector. This would mean only n bits would be required in the state vector for this process. The table will require extra memory but this will be offset when a large enough number of states is visited. Thus the only penalty will be in the run-time since the lookup in the table must be performed for every state generated.

In general, consider the state vector, s, of length n to be constructed of $p+1$ parts each of length n_i with $i = 0 \ldots p$. The parts s_i might be the bits allocated to a process, a variable or even a byte boundary. For every part s_i allocate a table t_i with k_i entries. Let the compressed state, c, be of length m also consisting of $p+1$ parts each of length $m_i = log_2 k_i$. When a new state $s = s_0 s_1 \ldots s_p$ is generated take part s_i and check if these bits are in the table t_i, if so, take the table index and assign it to c_i; if s_i is not in the table find the next open slot (starting from slot 0), insert s_i, and assign the index to c_i. The compression technique is illustrated in figure 9. When will this compression fail to save memory? If there are too many unique s_i parts then the table t_i will become too large, or if a fixed table size is assumed, the table will overflow. Thus this compression method is only suitable for models where each s_i part only has $k \ll 2^{n_i}$ unique states. There is also a break even point in the number of states that need to be generated before the memory used for the tables is offset. This value can be calculated as follows:

Let $|S|$ be the number of states generated. Memory will be saved when the memory used without compression is greater than the memory required if compression is used. Therefore,

$$|S|\sum_{i=0}^{p} n_i > |S|\sum_{i=0}^{p} log_2 k_i + \sum_{i=0}^{p} m_i n_i$$

and thus if,

$$|S| > \frac{\sum_{i=0}^{p} m_i n_i}{\sum_{i=0}^{p} n_i - \sum_{i=0}^{p} log_2 k_i}$$

This is read as: the number of states visited must be larger than the number of bits in the tables divided by the number of bits saved in each state. This fits in nicely with the intuition that the method will only work well if the tables are relatively empty, i.e. each s_i part only has a few unique states.

The implementation used here is to divide the state vector into 4 byte parts, i.e. every n_i is 4 bytes long. The table sizes for all t_i is fixed at 256, i.e. k_i is 256. Thus every s_i will be compressed into one byte in the compressed state. This configuration was chosen due to its simplistic implementation in the SPIN tool. The memory required by one table will therefore be $256 * 4$ bytes (1 kilobyte). Thus if the state vector is n bytes (note in SPIN the state vector is always a multiple of 8 bits) long the tables will require $\lceil \frac{n}{4} \rceil$ kilobytes of memory.

5.1. Related Work. Independently of the work reported here the compression technique described above was implemented in the SPIN system by Gerard Holzmann. This implementation, referred to as collapsing the state vector, is more general than the compression of 4 bytes into one described above (which will be called compression in the sequel). It allows the collapsing of each process into either 1, 2, 3 or 4 bytes, depending on how many unique states the largest process generates. Similarly all global variables are grouped together and collapsed. Since PROMELA (the language used to describe the models to be checked by SPIN) supports asynchronous communication, every channel (queue) is also collapsed in a similar fashion to each process. The collapsing of each state therefore consists of three parts: globals (as a unit), processes (one or more) and channels (zero or more). It was found that each process (and globals) in snoopy, pftp and scheduler can be collapsed into 2 bytes, and, the processes (and globals) in the address interface were small enough to be collapsed into 1 byte.

In table 2 the comparative results between collapsing and compression are shown when used in conjunction with the hash table store. For each method the memory (Mb), time (secs) and the number of bytes in the reduced state are shown. As expected the memory usage shows a four-fold reduction when using compression.

Model	Compress			Collapse			New Collapse		
	Mem	Time	Bytes	Mem	Time	Bytes	Mem	Time	Bytes
addr interface	0.17	4.1	19	0.25	5.4	27	0.11	2.3	12
snoopy	1.92	29	22	4.0	33	46	1.57	23.3	18
scheduler	3.67	221	15	7.8	234	32	4.41	226	18
pftp	12	741	29	10.9	730	26	5.8	659	14

TABLE 2. Results with state compression added to SPIN.

Significantly, though, the time increase is only 1.46 in the worst case. Note that the concurrent counter models cannot be handled by this method, since the 4 byte parts generate more than 256 unique states and therefore cause the tables to overflow. They were therefore not considered in the comparisons. Unexpectedly, the collapse method only performed better in reducing the state size for the pftp model. On closer inspection it was found that redundant information was stored in the state vector when channels were globally defined (as is the case in all four examples). Specifically, the channel contents, was stored in the "globals" section of the reduced state as well as in the "channels" section. The collapse implementation was changed such that no global channel will be stored in the "channels" section of the reduced state. This caused a considerable reduction in the state size as can be seen from the last column in table 2. Now for all but the scheduler example the collapse method saves memory. A further optimisation to the collapse method would be to allow the user to specify the number of bytes each process will require in the reduced state. Currently, the largest process, globals or channels determine how many bytes will be used during collapsing of the state vector. For example, in the scheduler only the globals (281 states) and one of the processes (274 states) require 2 bytes and all the other processes generate less than 256 states and can therefore be collapsed into 1 byte. Therefore, the reduced state of the scheduler could be 12 bytes instead of the current 18. As is the case with compression the time increase incurred is not significant.

5.2. Combining Compression with the OBDD State Store. In the previous section it was shown how the state vector length can be reduced. The effect of this reduction on the performance of the OBDD state store is now investigated. Note however that any reduction technique can be combined with the OBDD state store, even probabilistic ones. The results of this experiment are shown in table 3. The results indicate that the compression scheme now makes the OBDD state store a much more viable option. For instance, the pftp model that took more than 16 hours to generate only half its reachable states when just the OBDD state store was used can now complete the search in a reasonable time when compression is added. When considering the memory required for the normal hash table state store (table 1) it is evident that the OBDD state store combined with state compression is more space efficient for all four of the examples. The best reduction is achieved for the scheduler where 30 times less memory is required.

Interestingly, in the snoopy and scheduler models the compression method that yielded the best reduction in the size of the state vector did not save the most memory when the OBDD state store was added. However, in both these cases the

	OBDD + Compress			OBDD + Collapse		
Model	Memory	Nodes	Time	Memory	Nodes	Time
addr interface	0.29	11246	340	0.2	6950	212
snoopy	1.97	108193	3910	2.6	126141	4613
scheduler	0.49	19705	6891	0.48	18474	6185
pftp	6.3	349210	17403	3.43	167767	15870

TABLE 3. OBDD results with state compression/collapsing added to SPIN.

difference in state size is small and the results just highlights the unpredictability of the growth of OBDDs. In the pftp example, where the difference in state size is significant, the method which yielded the smaller state vector (collapse) also required less nodes in its largest OBDD.

Unfortunately, time requirements are now becoming an increasing problem. It would seem that the space savings are almost directly proportional in the time increase. This would seem to indicate that this method of combining an OBDD state store and state compression will only be suitable where memory is a bottleneck. Until now, however, we have only considered models where the complete state space is traversed. With the use of partial order methods it might be unnecessary to traverse all interleavings of transitions in concurrent processes. Using partial orders in combination with the OBDD state store is investigated in the next section.

6. Partial Orders

Partial orders are not discussed in any great detail here, the interested reader is referred to [**Val90, GW93, Pel94, HP94**]. It is sufficient to know that when partial orders are exploited during state generation it might cause certain states never to be generated. This is due to the fact that certain interleavings of independent transitions might not be considered. In [**HP94**] the implementation of partial order rules in the SPIN system is described.

When combining the OBDD state store (with the collapse method of state compression) with the partial order rules the results were surprising (see table 4). The expected reduction in memory is less than might have been anticipated, especially considering so many fewer states are visited. In fact, when considering the second last column in table 4, showing the memory required if partial orders and compression were used with the traditional hash table (i.e. not using an OBDD state store), it can be seen that using the hash table is more memory efficient than the OBDD state store when used in combination with partial orders.

In figure 10 the number of nodes in the OBDD state store is plotted during state generation for our four examples (with and without partial order rules). For pftp (top left) the OBDD grows faster when partial order rules are used. Even more surprising is the graph for the address interface: here the same number of states are visited with and without partial order rules, but the OBDD state store requires more nodes when partial orders are enforced! The reason for this behaviour can be explained by remembering that OBDDs perform well when there is regularity

Model	OBDD + Partial Orders				Partial Orders + Collapse	
	States	Memory	Time	Nodes	Memory	Time
addr interface	9491	0.21	242	7229	0.11	2.3
snoopy	15295	0.9	171	44543	0.26	3.1
scheduler	9251	0.29	259	12619	0.13	1.1
pftp	95241	1.72	1325	90057	1.27	19.8

TABLE 4. OBDD state store with Partial Orders

in the state space, when partial order rules are used some of this regularity may be removed. This explains the behaviour seen for pftp. In the case of the address interface, however, it is simply that the states were generated in a different order when partial orders were used and this resulted in a loss of regularity in the state space.

7. Conclusion

We have investigated a new approach to reduce the memory requirements during on-the-fly model checking. An OBDD encoding of the states visited during the depth-first exploration of the state space was used to achieve this goal. The initial results (table 1) were not encouraging. Therefore, a state compression technique was developed that would reduce the length of the state vector and therefore the memory required during the generation of the state space. This compression method works well for models that generate a large number of states $k \ll 2^n$ where n is the length of the state vector. Although the method was introduced to improve the performance of the OBDD state store, it can be used with any type of state storage method. For instance, combining a specific implementation of the state compression with a hash table as a state store produced a four-fold memory reduction while only increasing the time taken by less than a factor of two compared to the case where no compression was used (see table 2). The more general implementation of the compression technique that is currently used in SPIN, was also considered and found to store unnecessary information in the compressed state in certain cases. The improved version of this so-called collapse method was found to achieve more than a four-fold reduction in the majority of the examples used. Combining the state compression and the OBDD state store gave reductions in the memory requirements over both the cases where compression and a hash table were used and an OBDD state store without compression (table 3). Lastly, the OBDD state store (with state compression) was combined with a partial order reduction method. Although the partial order rules did reduce the number of reachable states, the expected memory reductions did not occur. Due to the removal of certain interleavings during partial order reductions, and therefore certain previously reachable states, the size of the OBDD state store were not reduced as much as before by the three OBDD transformation rules. Therefore, when reducing the state space according to partial order rules, it is in general more space and time efficient to use a hash table state store (see table 4).

In summary, an OBDD state store must in general be used with some form of compression, even probabilistic compression, if it is to be a worthwhile mechanism

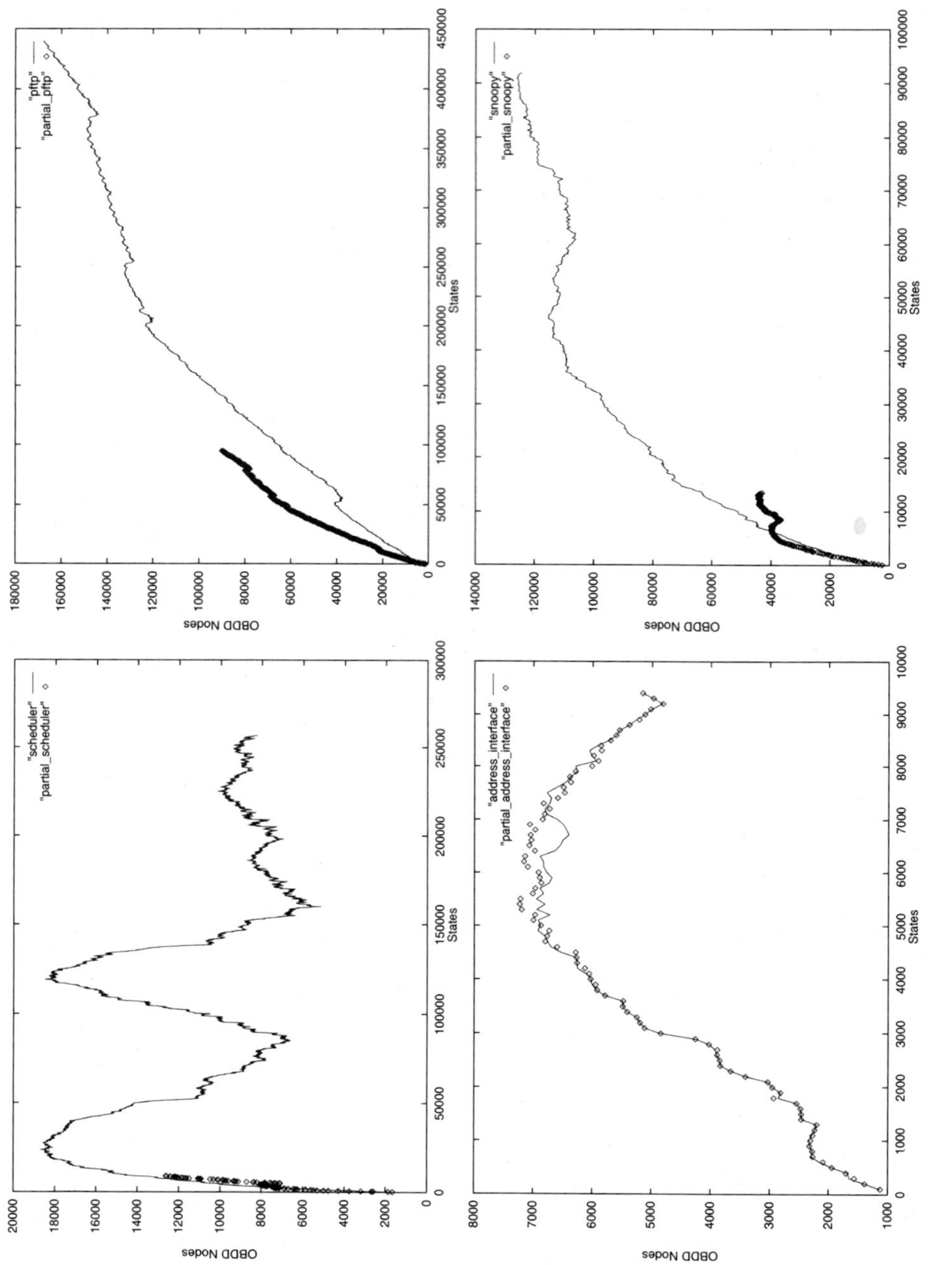

FIGURE 10. OBDD growth with and without partial order rules.

for space saving in SPIN. For all four of the real-life examples chosen here a space saving was achieved over the traditional hash table storage method when using the OBDD state store with compression. It must be said that a substantial time increase is incurred, but this in part due to the way the operations on the OBDDs

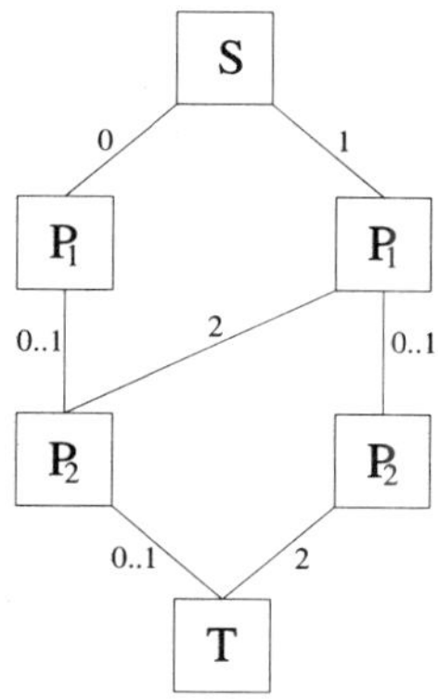

FIGURE 11. ONDD for the mutual exclusion system. Nodes are labelled with the three process variables.

were implemented. The operations are by no means time efficient since this was undertaken as an experiment in memory reduction. Furthermore, no mention is made of the performance of other variable orderings, but this is again because of the fact that we were interested in the general behaviour of an OBDD state store and were therefore not concerned about fine-tuning the method to obtain better performance. Improving the performance of the three OBDD functions used and experimenting with heuristics for finding better variable orderings will be the focus of future work.

Another future extension is to use ordered n-ary decision diagrams (ONDDs) [**Rau95**] instead of OBDDs. The main difference between ONDDs and OBDDs is that an ONDD node is n-ary, where n is the cardinality of the domain of the variable it is labelled with. Every node in an ONDD can therefore have n possible arcs to the nodes below it in the graph. Since it is possible that many of the arcs from a node can go to the same node, we will adopt the compaction described in [**Rau95**] where every arc is labelled by a range of constants rather than with an individual constant. In figure 11 the ONDD encoding the reachable states in the mutual exclusion example is shown. Note that the state compression scheme of section 5 will fit nicely with the use of an ONDD state store. For instance, every component c_i of the compressed state (see figure 9) will be one of the variables of the ONDD. Furthermore, if the indices of the tables t_i are allocated consecutively (as is the case in all the examples in this paper), then the chances of compacting the arcs from a node by labelling it with ranges are increased.

References

[BCG95] G. Bhat, R. Cleaveland, and O. Grumberg. Efficient On-the-fly Model Checking for CTL*. In *Symposium on Logic in Computer Science*, June 1995.

[BCM+90] J.R. Burch, E.M. Clarke, K.L McMillan, D.L. Dill, and L.J. Hwang. Symbolic Model Checking: 10^{20} States and Beyond. In *Proceedings of the 5-th IEEE Symposium on Logic in Computer Science*, pages 428–439, Philadelphia, June 1990.

[Bry86] R.E. Bryant. Graph-based Algorithms for Boolean Function Manipulation. *IEEE Transactions on Computers*, C-35(8):677–691, August 1986.

[Bry92] R. E. Bryant. Symbolic Boolean Manipulation with Ordered Binary-Decision Diagrams. *ACM Computing Surveys*, 24(3):293–318, September 1992.

[BVW94] O. Bernholtz, M. Vardi, and P. Wolper. An Automata-Theoretic Approach to branching-Time Model Checking. In *CAV '94 : 6th International Conference on Computer Aided Verification*, volume 818 of *Lecture Notes in Computer Science*, 1994.

[CES83] E.M. Clarke, E.A. Emerson, and A.P. Sistla. Automatic Verification of Finite State Concurrent Systems Using Temporal Logic Specifications: A Practical Approach. *Proceedings 10th ACM Symposium on Principles of Programming Languages*, pages 117–126, 1983.

[CES86] E.M. Clarke, E.A. Emerson, and A.P. Sistla. Automatic Verification of Finite-State Concurrent Systems Using Temporal Logic Specifications. *ACM Transactions on Programming Languages and Systems*, 8(2):244–263, April 1986.

[CGL93] E. Clarke, O. Grumberg, and D. Long. Verification Tools for Finite-State Concurrent Systems. In *A Decade of Concurrency: Reflections and Perspectives*, volume 803 of *Lecture Notes in Computer Science*, 1993.

[FDG+93] S. B. Furber, P. Day, J. D. Garside, N. C. Paver, and J. V. Woods. A Micropipelined ARM. In T. Yanagawa and P. A. Ivey, editors, *Proceedings of VLSI 93*, pages 5.4.1–5.4.10, September 1993.

[FdV91] W. Fouché and P. de Villiers. A Reusable Kernel for the Development of Control Software. In *6th Southern African Computer Symposium*, pages 83–94, July 1991.

[GHP92] P. Godefroid, G.J. Holzmann, and D. Pirottin. State Space Caching Revisited. In *Proc. 4th Computer Aided Verification Workshop*, Montreal, Canada, June 1992. also in: Formal Methods in System Design, Kluwer, 1995.

[Gre96] J-C. Gregoire. State Space Compression in SPIN with GETSs. In *Participants Proceedings of the 2nd International Workshop on the* SPIN *Verification System*, New Brunswick, USA, 5 August 1996.

[GW93] P. Godefroid and P. Wolper. Using Partial Orders for the Efficient Verification of Deadlock Freedom and Safety Properties. *Formal Methods in System Design*, 2(2):149–164, April 1993.

[Hol88] G.J. Holzmann. An Improved Protocol Reachability Analysis Technique. *Software, Practice & Experience*, 18(2):137–161, February 1988.

[Hol91] G.J. Holzmann. *Design and Validation of Computer Protocols.* Prentice-Hall, Englewood Cliffs, New Jersey, 1991.

[Hol95] G.J. Holzmann. An Analysis of Bitstate Hashing. In *15-th International Symposium on Protocol Specification, Testing, and Verification*, Warsaw,Poland, June 1995. North-Holland.

[HP94] G.J. Holzmann and Doron Peled. An Improvement in Formal Verification. In *Proc. FORTE94*, Berne, Switzerland, October 1994.

[McM92] K.L. McMillan. *Symbolic Model Checking.* PhD thesis, Carnegie Mellon University, May 1992.

[Pel94] D. Peled. Combining Partial Order Reductions with On-the-fly Model Checking. In *CAV '94 : 6th International Conference on Computer Aided Verification*, volume 818 of *Lecture Notes in Computer Science*, 1994.

[Rau95] A. Rauzy. Toupie = μ-calculus + constraints. In *CAV '95 : 5th International Conference on Computer Aided Verification*, volume 939 of *Lecture Notes in Computer Science*, 1995.

[SD95] U. Stern and D. Dill. Improved Probilistic Verification by Hash Compaction. In *CHARME '95 : Correct Hardware Design and Verification Methods*, volume 987 of *Lecture Notes in Computer Science*, 1995.

[Val90] A. Valmari. A Stubborn Attack on State Explosion. In *Proceedings of the Second International Conference for Computer-Aided Verification.* Lecture Notes in Computer Science, 531, June 1990.

[Vis93] W.C. Visser. A Run-Time Environment for a Validation Language. Master's thesis, Department of Computer Science, University of Stellenbosch, Stellenbosch 7600, South Africa, December 1993.

[VW86] M.Y. Vardi and P. Wolper. An Automata Theoretic Approach to Automatic Program Verification. In *First Symposium on Logic in Computer Science*, pages 322–331, June 1986.

[WL93] P. Wolper and D. Leroy. Reliable Hashing without Collision Detection. In *CAV '93 : 5th International Conference on Computer Aided Verification*, volume 697 of *Lecture Notes in Computer Science*, 1993.

[Wol87] P. Wolper. On the Relation of Programs and Computations to Models of Temporal Logic. In *Temporal Logic in Specification*, volume 398 of *Lecture Notes in Computer Science*, 1987.

DEPARTMENT OF COMPUTER SCIENCE, UNIVERSITY OF MANCHESTER, MANCHESTER, M13 9PL, UK

E-mail address: visserw@cs.man.ac.uk

DEPARTMENT OF COMPUTER SCIENCE, UNIVERSITY OF MANCHESTER, MANCHESTER, M13 9PL, UK

E-mail address: howard@cs.man.ac.uk

Selected Titles in This Series

(*Continued from the front of this publication*)

15 **Nathaniel Dean and Gregory E. Shannon, Editors**
Computational Support for Discrete Mathematics

14 **Robert Calderbank, G. David Forney, Jr., and Nader Moayeri, Editors**
Coding and Quantization: DIMACS/IEEE Workshop

13 **Jin-Yi Cai, Editor**
Advances in Computational Complexity Theory

12 **David S. Johnson and Catherine C. McGeoch, Editors**
Network Flows and Matching: First DIMACS Implementation Challenge

11 **Larry Finkelstein and William M. Kantor, Editors**
Groups and Computation

10 **Joel Friedman, Editor**
Expanding Graphs

9 **William T. Trotter, Editor**
Planar Graphs

8 **Simon Gindikin, Editor**
Mathematical Methods of Analysis of Biopolymer Sequences

7 **Lyle A. McGeoch and Daniel D. Sleator, Editors**
On-Line Algorithms

6 **Jacob E. Goodman, Richard Pollack, and William Steiger, Editors**
Discrete
and Computational Geometry: Papers from the DIMACS Special Year

5 **Frank Hwang, Fred Roberts, and Clyde Monma, Editors**
Reliability of Computer and Communication Networks

4 **Peter Gritzmann and Bernd Sturmfels, Editors**
Applied Geometry and Discrete Mathematics, The Victor Klee Festschrift

3 **E. M. Clarke and R. P. Kurshan, Editors**
Computer-Aided Verification '90

2 **Joan Feigenbaum and Michael Merritt, Editors**
Distributed Computing and Cryptography

1 **William Cook and Paul D. Seymour, Editors**
Polyhedral Combinatorics